Semiconductor Essentials

A Leader's Express reference to Electronics Concepts

By

Rupal Jain

Copyright © 2024 by Rupal Jain

All rights reserved. No part of this publication may be reproduced, distributed, or transmitted in any form or by any means, including photocopying, recording, or other electronic or mechanical methods, without the prior written permission of the publisher, except in the case of brief quotations embodied in critical reviews and certain other noncommercial uses permitted by copyright law.

About the Book

In the realm of semiconductor devices, navigating through the vast sea of concepts and equations can often feel like an overwhelming journey. While there are numerous books available on the subject, many readers find themselves longing for a concise handbook that provides quick access to fundamental principles and advanced equations without the need to wade through extensive explanations.

This book aims to fill that gap.

Tailored for experienced engineers, industry leaders in semiconductor manufacturing, and professionals involved in chip-making and disciplines related to Electrical and Electronics, this book serves as a comprehensive reference, condensing essential topics and equations critical for device concepts. semiconductor development, and design into easily accessible formats. Whether you're seeking quick access to support Research and Innovation in cross-disciplinary fields such as Electronics, Applied Physics, Photonics, Material Science, refreshing your knowledge of semiconductor fundamentals, revisiting concepts for chip failure analysis, IC design, failure mode effect analysis, or preparing to be a leader with a broad and structured understanding of device concepts, this meticulously crafted book is designed to meet your diverse needs with precision and efficiency.

From basic semiconductor theory to advanced equations, each section is meticulously crafted to provide clarity and accessibility without sacrificing depth. With a focus on brevity and precision, this book empowers readers to swiftly locate the information they seek, making it an indispensable companion for anyone working in the field of semiconductor devices.

Additionally, this book is intended to prepare leaders in chip manufacturing, research, and innovation across various disciplines including electrical electronics, photonics, material sciences, and related fields. Regarding the production process, this book has undergone several critical stages, including writing, peer reviews, proofreading, and feedback incorporation, all within a comprehensive quality management system.

The author has diligently strived to ensure the book is error-free to the best of their ability. However, if any mistakes have inadvertently occurred, the author sincerely regrets them and would greatly appreciate anyone bringing them to their attention. It's important to note that the author does not assume legal responsibility, and shall not be liable for any direct, indirect, incidental, consequential, or punitive damages arising from the use of or inability to use the content of this book.

As you embark on your journey through this book, may it serve as a reliable companion, providing you with the knowledge and insights you need to navigate the complexities of semiconductor devices, with confidence and ease!

Happy reading!

Rupal Jain

(Author)

Semiconductor Essentials: A Leader's Express reference to Electronics Concepts

About the Author

Rupal Jain is a distinguished figure in the field of semiconductor chip manufacturing, with extensive expertise in engineering, program management, and strategic alignment. Throughout her career, she has spearheaded projects encompassing the entire chip development lifecycle – from design conception and quality management to global delivery across regions like the USA, Taiwan, Singapore, Italy, Malaysia, China, and India. Her profound knowledge is recognized by prestigious certifications like PMP, CSM, and Lean Six Sigma Black Belt. Rupal holds a master's degree in electrical and Electronics Engineering, earned through a joint program between NTU Singapore and TUM Germany.

Beyond her technical prowess, Rupal's innovative contributions have garnered international acclaim. She is a frequent contributor to esteemed publications, serves on prestigious juries, and holds nominated memberships in industry organizations. Notably, her work has been recognized with coveted awards and patents, further solidifying her position as a leader in the field. Her upcoming books, "Mastering Project Management: PMP and Agile for Leaders" and "Advancements in AI and IoT for Chip Manufacturing and Defect Prevention," promise to share her valuable insights with the next generation of leaders and engineers.

Table of Contents

About the Book   i

About the Author   iii

CHAPTER-1 INTRODUCTION TO SEMICONDUCTORS   1

1.1 Models and Equations for semiconductor device Physics   2

1.2 Doping an n-type substrate with p-type impurities.   5

1.3 Comparison of Energy Band Diagrams in n-type, p-type semiconductors   6

1.4 Effect of Temperature on Carrier Concentration   8

1.5 Effect of Temperature on Fermi Level   11

1.6 Mean Free Path, Electric Field, mobility, Temperature, and Current   13
    1.6.1 Drift-Diffusion Model   16

1.7 Semiconductor in Equilibrium vs. Non-Equilibrium   17

1.8 Resistivity of Semiconductors   20

1.9 Diffusion Length and Equation   22

1.10 The Einstein Relation- Are Diffusion Coefficient and mobility independent?   24

1.11 Band Bending   25

1.12 Electrical Analogy of Scattering in Semiconductors   27

1.13 Optical Processes in Semiconductors: Absorption and Emission   28

1.14 Ambipolar Transport in Semiconductors   30

1.15 Hall Effect: Explanation with Equations and Diagrams	31

CHAPTER-2 P-N DIODE   **34**

2.1 Introduction to P-N Diode	34
2.1.1 Diode Current Equation (Shockley Diode Equation)	36
2.1.2 Diode Voltage Equation (Ideal Diode Equation)	37
2.1.3 Diode Resistance	37
2.1.4 Diode Capacitance	37
2.1.5 Space Charge Width	38
2.2 Reverse Bias with P-N Diode	39
2.2.1 Difference between Junction Capacitance and Diode capacitance	41
2.3 Forward Bias with P-N Diode	43
2.3.1 Excess Carrier Concentration	44
2.3.2 Forward Bias Resistance, capacitance, and admittance	45
2.3.2.1 Resistance	45
2.3.2.2 Capacitance	46
2.3.2.3 Admittance	47
2.4 I-V characteristics of a Diode	48
2.5 Comparison between P-N Junction characteristics for Zero Bias, Reverse Bias, Forward Bias	51
2.6 Junction Breakdown	51
2.7 Special types of PN JUNCTION DIODE	53
2.7.1 Tunnel Diode	53
1. Basic Functionality	53
2. Tunneling Effect	53
3. Current-Voltage Characteristic	53
4. Applications	54
5. Limitations	54
6. Tunnel Diode Equivalent Circuit	55

2.7.2 PIN Diode 55
    1. Key Characteristics 56
    2. Applications 56
    3. Advantages of PIN Diodes 56

2.7.3 Varactor Diode 57
    1. How Does It Work? 57
    2. Characteristics 57
    3. Typical Applications 58

2.7.4 Schottky Diode 58
    1. Key characteristics 58
    2. Applications 59
    3. Comparison to standard p-n junction diodes 59

CHAPTER-3 BIPOLAR JUNCTION TRANSISTOR (BJT) 61

3.1 What is a Transistor? 61

3.2 Types of Transistors 61
    3.2.1 Bipolar Junction Transistors (BJTs) 61
    3.2.2 Field-Effect Transistors (FETs) 62
        1. Subtypes of FETs 62

3.3 Basic Structure of a Bipolar Junction Transistor (BJT) 63

3.4 Typical Doping Concentrations for BJT 64

3.5 Transistor Biasing 66

3.6 Transistor Operation 68

3.7 Current Components in a Transistor 70

3.8 Transistor characterization 71
    3.8.1. Circuit-Level Analysis with Detailed BJT Models 74
        Case Study 1: Common-Emitter Amplifier with Ebers-Moll Model 74

Case Study 2: High-Frequency BJT Amplifier with Gummel-Poon Model
                                                                      75

3.9 Transistors at low frequency: Characteristics and Analysis   76
   3.9.1 Hybrid Model for CB Configurations   77
   3.9.2 Hybrid Model for CE Configuration   78
   3.9.3 Hybrid Model for CC Configuration   79
   3.9.4 Questions based on Design and Simulation   80
      3.9.4.1 Circuit Design and Analysis   80
      3.9.4.2. Advanced Modeling and Analysis   81
      3.9.4.3 Real-World Applications and Troubleshooting   82
      3.9.4.4 Leadership and Problem-Solving   83

3.10 Transistors at high frequency   84
   3.10.1 Validity and Parameter Variation   88
   3.10.2 Current gain High frequency BJT vs Low frequency BJT   89
   3.10.3 Current gain with resistive load   91
   3.10.4 Frequency response of single stage CE Amplifier   92
   3.10.5 Definition of GAIN- BANDWIDTH PRODUCT   93

3.11 Applications   94
   3.11.1 Rectifiers   94
   3.11.2 Filters   96
   3.11.3 Feedback Amplifiers   98
   3.11.4 Oscillators   99
   3.11.5 Multistage Amplifiers   102

CHAPTER-4 FIELD EFFECT TRANSISTORS/ FETS   **105**

4.1 Introduction to MOS structure as two terminal capacitors   105

4.2.1 Energy Band for MOS Capacitors with p-type substrate   107

4.2.2 Energy Band for MOS Capacitors with n-type substrate   112

4.3 Types of FET and symbology   114

4.4 MOSFET Biasing and Configuration 115
    4.4.1 Models and Equations with Characterization 117

4.5 Comparison between FINFET, MOSFET, Replacement Gate, Floating Gate
    118
    4.5.1 Mathematical equations 120
    4.5.2 Capacitance-Voltage characteristics of a MOS Cap 122

4.6 Second order Effects in FETs 123
    4.6.1 Threshold Voltage Roll-Off 124
    4.6.2 Drain-Induced Barrier Lowering (DIBL) 124
    4.6.3 Hot Carrier Injection 124
    4.6.4 Subthreshold Slope Degradation 124

INDEX   **126**

REFERENCES   **129**

Chapter-1

Introduction to Semiconductors

Semiconductors play a pivotal role in modern electronics, serving as the foundation for various devices that have revolutionized our daily lives. At the heart of semiconductor technology lies the concept of equilibrium, a fundamental principle that governs the behavior of semiconductors under certain conditions. In this introduction, we delve into the intriguing world of semiconductors in equilibrium, exploring the underlying principles, characteristics, and implications of this essential concept[1].

Semiconductors are materials that possess unique electrical properties, making them distinct from conductors and insulators. Unlike conductors, which readily allow the flow of electrical charge, and insulators, which inhibit the flow of charge, semiconductors exhibit a conductivity that falls between these extremes. This intermediate conductivity is a result of the presence of energy bands within the semiconductor material, which determine the mobility of charge carriers such as electrons and holes[2].

In equilibrium, a semiconductor reaches a stable state where there is no net flow of charge carriers. This equilibrium state occurs when the rate of carrier generation equals the rate of carrier recombination, resulting in a balance between the concentration of electrons and holes within the semiconductor material. Understanding semiconductor behavior in equilibrium is essential for predicting and analyzing the performance of semiconductor devices in practical applications.

In this exploration of semiconductors in equilibrium, we will delve into key concepts such as the energy band structure of semiconductors, the Fermi level, and the mechanisms governing carrier generation and recombination. We will also examine the influence of doping, temperature, and external bias on semiconductor behavior in equilibrium. Through this journey, we aim to gain a deeper appreciation for the intricate interplay of factors that shape the behavior of semiconductors and drive the advancement of modern technology[3].

1.1 Models and Equations for semiconductor device Physics

1. **Intrinsic Semiconductor Model**: In an intrinsic semiconductor, the number of electrons in the conduction band is equal to the number of holes in the valence band[3,4,5,6,8]. This equilibrium[7] condition can be represented mathematically by the equation:

$$n_i^2 = n \cdot p$$

where n_i is the intrinsic carrier concentration, n is the electron concentration, and p is the hole concentration.

2. **Extrinsic Semiconductor Model**: Extrinsic semiconductors[7,8,9,10] are doped with impurities to modify their electrical properties. The carrier concentration in an extrinsic semiconductor is given by the law of mass action, which incorporates doping concentrations. For an n-type semiconductor doped with donor impurities, the electron concentration (n) is given by:

$n = n_i^2 / N_D$ where N_D is the donor concentration.

For a p-type semiconductor doped with acceptor impurities[11,12,13], the hole concentration (p) is given by:

$p = n_i^2 / N_A$ where N_A is the acceptor concentration.

3. **Energy Band Diagram**: The energy band diagram represents the energy levels of electrons in a semiconductor. It's often described using the Schrödinger equation for electrons in a crystal lattice potential.

4. **Doping**: Doping is the process of intentionally introducing impurities into a semiconductor material to modify its electrical properties. Doping can increase the conductivity of the semiconductor and alter its carrier concentration.

5. **Carriers**: Carriers are charge carriers that transport electrical charge through a semiconductor. In semiconductors, carriers can be either electrons (negative charge) or holes (positive charge). The concentration of carriers in a semiconductor material determines its electrical conductivity.

6. **Carrier Properties**: Carriers in semiconductors exhibit properties such as mobility (μ), which represents the ease of carrier movement under an electric field. Carrier mobility is influenced by factors like doping concentration, temperature, and crystal structure. Another important property is the carrier lifetime (τ), which represents the average time a carrier remains mobile before recombining.

7. **Donor Impurities**: Donor impurities are atoms that introduce excess electrons into the semiconductor crystal lattice. Common donor impurities in silicon include phosphorus (P) and arsenic (As). Donor atoms have an extra valence electron compared to the semiconductor atoms, making them easily ionizable.

8. **n-Type Semiconductor**: An n-type semiconductor is doped with donor impurities, resulting in an excess of free electrons as majority carriers. The electron concentration (n) in an n-type semiconductor is much higher than the hole concentration (p).

9. **p-Type Semiconductor**: A p-type semiconductor is doped with acceptor impurities, resulting in an excess of holes as majority carriers. The hole concentration (p) in a p-type semiconductor is much higher than the electron concentration (n).

10. **Carrier Concentration**:
    For an n-type semiconductor: $n = N_D$
    For a p-type semiconductor: $p = N_A$

    where N_D is the donor concentration and N_A is the acceptor concentration.

11. Formation of N type and P type semiconductors

- **n-Type Semiconductor**: Doping with Phosphorus adds extra electrons to the crystal lattice, creating an excess of negative charge carriers (electrons). Phosphorus acts as a donor impurity, introducing energy levels near the conduction band of the semiconductor. This lowers the energy required for electrons to move into the conduction band, increasing conductivity[14,15,16].
- **p-Type Semiconductor**: Doping with Boron creates "holes" in the crystal lattice where electrons are missing, effectively introducing positive charge carriers. Boron acts as an acceptor impurity, creating energy levels near the valence band. This lowers the energy required for electrons to jump from the valence band to the acceptor energy level (hole), increasing conductivity[17].

12. Compensated Semiconductor- A compensated semiconductor is a type of semiconductor material where the number of dopant atoms added to the crystal lattice is balanced in such a way that the material maintains overall electrical neutrality. In other words, the concentration of donor impurities (which contribute extra electrons) is roughly equal to the concentration of acceptor impurities[18] (which create "holes" where electrons are missing).

In compensated semiconductors, the donor and acceptor impurities compensate each other's effects, resulting in a material that is not strongly doped either as an n-type or p-type semiconductor. This balancing act allows for precise control over the electrical properties of the semiconductor, making it useful in certain applications where specific electrical characteristics are required.

Compensated semiconductors are often employed in devices where precise tuning of conductivity is necessary, such as in some types of sensors, photovoltaic cells, and certain integrated circuits.

1.2 Doping an n-type substrate with p-type impurities.

Doping an n-type substrate with p-type impurities introduce competition between the majority carriers (electrons in n-type) and the minority carriers (holes in p-type). The specific outcome depends on the relative doping concentrations[22]:

1. **Light Doping:** If the p-type dopant concentration is **significantly lower** than the n-type dopant concentration, the overall conductivity remains **n-type**, but with slightly reduced electron mobility due to scattering with the introduced holes. This scenario is often referred to as **lightly doped p-type in n-type (p+ in n)** and can be used to create shallow p-n junctions for various electronic devices[19].

2. **Heavy Doping:** If the p-type dopant concentration is **comparable to or higher** than the n-type dopant concentration, the material can undergo a type conversion. This means the majority carrier type changes from electrons to holes, and the overall conductivity becomes **p-type**. This scenario is referred to as **heavily doped p-type in n-type (p++ in n)** and can be used to create specific device regions with desired conductivity profiles[21].

Similarly, you can derive what would happen if P type is doped with N type impurities.

Characteristics	n-type Semiconductor	p-type Semiconductor
Type of Dopant	Donor impurity (e.g., Phosphorus in Silicon)	Acceptor impurity (e.g., Boron in Silicon)
Effect on Conduction Band	Raises the energy level of the conduction band (CB) electrons	Lowers the energy level of the CB electrons
Electron Concentration	Higher electron concentration due to excess donor electrons	Lower electron concentration due to fewer free electrons

Semiconductor Essentials

Characteristics	n-type Semiconductor	p-type Semiconductor
Hole Concentration	Lower hole concentration due to fewer holes	Higher hole concentration due to excess acceptor impurities
Electron Mobility	High electron mobility due to greater number of free electrons	Lower electron mobility due to fewer free electrons
Hole Mobility	Lower hole mobility due to fewer holes	High hole mobility due to greater number of holes

Table 1 Characteristics of N-type, P-type semiconductors

1.3 Comparison of Energy Band Diagrams in n-type, p-type semiconductors

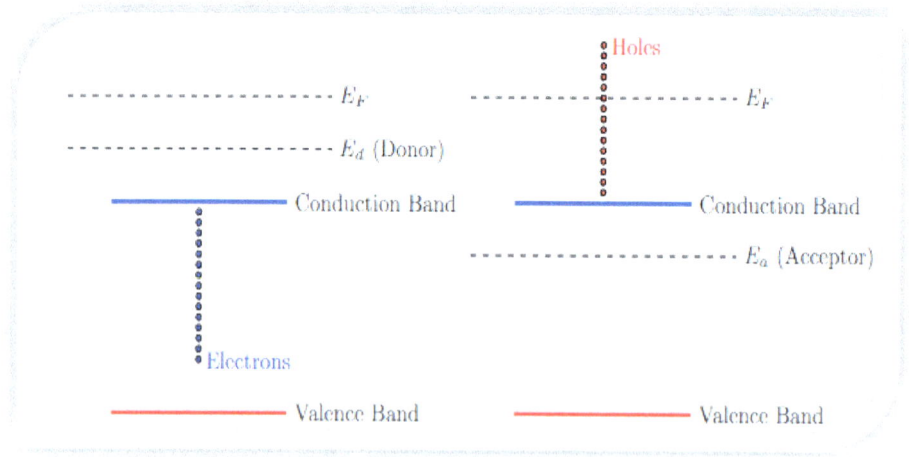

Figure 1 Energy Band Diagrams of N-type, P-type semiconductors.

Eg: **Bandgap energy (eV);** *Ec*: **Conduction band minimum energy (eV);** *Ev*: **Valence band maximum energy (eV)**

1. N-Type Semiconductor:
    - Conduction Band: Contains mobile electrons.
    - Valence Band: Occupied by electrons in their ground state.
    - Fermi Level (E_F): Close to the conduction band due to excess electrons.
    - Doping Agent: Typically doped with Group 15 elements (e.g., Phosphorus).
2. P-Type Semiconductor:
    - Conduction Band: Empty or sparsely populated.
    - Valence Band: Contains holes due to excess positive charge carriers.
    - Fermi Level (E_F): Close to the valence band due to excess holes.
    - Doping Agent: Typically doped with Group 13 elements (e.g., Boron).
3. Metal:
    - Conduction Band: Overlaps with the valence band, forming a continuous band.
    - Valence Band: Partially filled with electrons.
    - Fermi Level (E_F): Lies within the conduction band due to the high electron density.
    - Electrical Conductivity: High due to the presence of free electrons.
4. Insulator:
    - Conduction Band: Empty or significantly separated from the valence band.
    - Valence Band: Occupied by electrons in their ground state.
    - Fermi Level (E_F): Close to the valence band, as few electrons are excited to the conduction band.
    - Electrical Conductivity: Very low due to the large band gap.

Comparison:

- In semiconductors, the Fermi level lies close to either the conduction or valence band, depending on the type (n-type or p-type).
- Metals have overlapping conduction and valence bands, with the Fermi level within the conduction band.
- Insulators have a large band gap, with the Fermi level close to the valence band, limiting electron mobility[20].

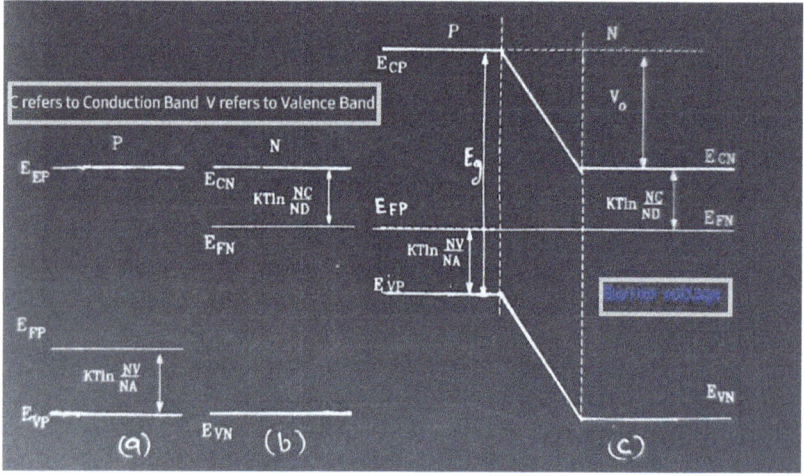

Figure 2 Energy Band Diagrams of N-type, P-type semiconductors with equations and barrier voltage

1.4 Effect of Temperature on Carrier Concentration

For a n type semiconductor

$$E_F = E_i - \frac{kT}{q} \ln\left(\frac{N_d}{N_c}\right)$$

$$n = N_c \exp\left(\frac{E_F - E_c}{kT/q}\right)$$

For a p type semiconductor

$$E_F = E_i + \frac{kT}{q} \ln\left(\frac{N_a}{N_v}\right)$$

$$p = N_v \exp\left(\frac{E_v - E_F}{kT/q}\right)$$

- EF represents the Fermi level energy (eV).
- Ei represents the intrinsic Fermi level energy (eV).
- Ec represents the Fermi level energy (eV).
- k represents the Boltzmann constant (J/K).
- T represents the temperature (K).
- Na represents the acceptor concentration (cm^-3).
- Nv represents the effective density of states in the valence band (cm^-3).
- n represents electron concentrations in cm 3
- p represents hole concentrations in per-cm 3
- q represents the electron charge (C).
- Nd represents the donor concentration (cm^-3).
- Nc represents the effective density of states in the conduction band (cm^-3).

Alternate representation at Fermi-Dirac Equation

$$f(E) = \frac{1}{1 + \exp\left(\frac{E - E_F}{kT/q}\right)}$$

where:

- *f(E)*: Probability of finding a fermion at energy E.
- E: Energy of the state (eV).
- E_F: Fermi level energy (eV), representing the energy level at which the probability of occupation is 50%.
- k: Boltzmann constant (J/K).
- T: Temperature (K).
- q: Electron charge (C).

Understanding the Equation: The exponential term determines the "steepness" of the distribution. A higher exponent lead to a steeper curve, indicating a sharper transition between occupied and unoccupied states.

The Fermi level (E_F) acts as a reference point. States below E_F have a higher probability of occupation, while states above E_F have a lower probability.

Temperature (T) plays a crucial role. Higher temperatures lead to a larger denominator in the exponent, making the distribution "broader" and increasing the probability of finding fermions at higher energy levels[21,26].

i. **Proportionality:**

- The Fermi-Dirac function, **f(E)**, doesn't directly give the exact probability of finding a fermion at a specific energy level[23].
- Instead, it represents the **relative probability** compared to a reference point.
- This reference point is the **Fermi level (E_F)**, where the probability of occupation is **50%**.

ii. **Probability:**

- While **f(E)** doesn't provide absolute probability, it's **proportional** to the actual probability.
- This means:
    - **f(E) = 1** implies a **high probability** (close to 100%) of finding the fermion at that energy.
    - **f(E) = 0** implies a **very low probability** (close to 0%) of finding the fermion at that energy.
    - Values between 0 and 1 represent **intermediate probabilities** based on their position relative to 1 (high probability) and 0 (low probability).

iii. **Analogy:** Imagine a light switch that controls a room's brightness.

- **f(E) = 1** is like turning the switch fully on, resulting in a bright room (high probability of finding the fermion)[25].
- **f(E) = 0** is like turning the switch fully off, resulting in a dark room (low probability of finding the fermion).
- Values between 0 and 1 represent dimming the light to various levels (varying probabilities).

Remember:

- The Fermi-Dirac function **f(E)** provides a **relative sense** of how likely it is to find a fermion at different energy levels, with the Fermi level serving as a reference point for this comparison.

1.5 Effect of Temperature on Fermi Level

The **Fermi level (μ)**, which represents the energy level at which the probability of finding an electron is 50%, is not directly proportional to temperature (T). However, temperature can indirectly influence the Fermi level through various mechanisms, depending on the material and its properties. Here's a breakdown of the potential effects[22]:

1. Metals:

- In **metals**, increasing temperature generally leads to a **decrease in the Fermi level**. This is because thermal vibrations of atoms increase the scattering of electrons, effectively reducing their mean free path and mobility. Consequently, the electrons occupy slightly lower energy states to maintain the same overall energy distribution according to the Fermi-Dirac statistics.
- However, the magnitude of this decrease is typically **small** compared to the absolute value of the Fermi level in metals, which lies several electron volts below the vacuum level [23].

2. Semiconductors:

- In **intrinsic semiconductors**, the effect of temperature on the Fermi level is more complex and depends on the bandgap and carrier concentration.
    - At **low temperatures**, the Fermi level typically lies near the middle of the bandgap.
    - As temperature **increases**, the intrinsic carrier concentration (both electrons and holes) increases due to thermal excitation across the bandgap. This can cause the Fermi level to **shift towards the majority carrier band (conduction band for n-type, valence band for p-type)** to maintain charge neutrality[24].
- In **doped semiconductors**, the presence of dopant atoms pins the Fermi level closer to their respective bands, making it less sensitive to temperature variations compared to intrinsic semiconductors.

3. Insulators:

- In **insulators**, the Fermi level typically lies deep within the bandgap, and temperature variations have a **minimal effect** on its position due to the large bandgap and negligible intrinsic carrier concentration.

Additional factors:

- The specific material properties, such as band structure, effective masses, and defect densities, can influence the exact nature of the temperature dependence of the Fermi level.
- External factors like applied electric fields and doping can further modify the Fermi level position and its response to temperature changes.

1.6 Mean Free Path, Electric Field, mobility, Temperature, and Current

1. Mean Free Path (λ):

- Represents the average distance a charge carrier (e.g., electron) travels between collisions with other particles in a material[24].
- **Equation:** $\lambda = v_{avg} * t$
  - v_{avg}: Average thermal velocity of the charge carriers (m/s).
  - t: Average time between collisions (s).

2. Electric Field (E):

- Applies a force on charged particles, causing them to accelerate.
- **Effect on Drift Velocity (v_d):** $v_d = \mu * E$, where:
  - μ: Mobility of the charge carriers (m^2/Vs).
  - This equation shows that a stronger electric field (higher E) leads to a higher drift velocity, indicating faster movement of the charged particles[1,2,3,4,5,6,7,8].

3. Temperature (T):

- Affects the thermal motion of charge carriers, influencing their average velocity and collision frequency.
- **Effect on Mobility (μ):** Generally, mobility decreases with increasing temperature due to more frequent collisions (except for specific cases in certain materials).

$$\mu(T) = \mu_0 \left(\frac{T}{T_0}\right)^{-\alpha}$$

where:

- $\mu(T)$: Mobility at temperature T

- μ_0: Reference mobility at a specific temperature (e.g., T_0)
- α: Material-dependent constant

4. Properties of mobility with Electric Field:

The **electric field** plays a crucial role in determining the **mobility (μ)** of charge carriers (electrons and holes) in semiconductors. Here's a breakdown of the key properties:

a. **Low-Field Regime:**

- At **low electric fields**, the relationship between electric field (E) and drift velocity (vd) is **linear**. This regime is often referred to as the **ohmic region**[23].

- In this region, carrier mobility (μ) remains **constant** and can be expressed as:

$$\mu = v_d / E$$

b. **High-Field Regime:** As the electric field **increases** beyond a certain threshold, the relationship between E and v_d becomes **non-linear**. This is due to various scattering mechanisms that become more prominent at higher fields:

- **Increased collisions:** Carriers experience more frequent collisions with phonons (lattice vibrations) and impurities, hindering their movement.

- **Optical phonon emission:** Carriers lose energy by emitting optical phonons, reducing their average velocity.

c. **Saturation Velocity:** At **very high electric fields**, the drift velocity **approaches a constant value** called the **saturation velocity (v_{sat})**. This is because even increasing the electric field further cannot overcome the limitations imposed by scattering mechanisms[11,12].

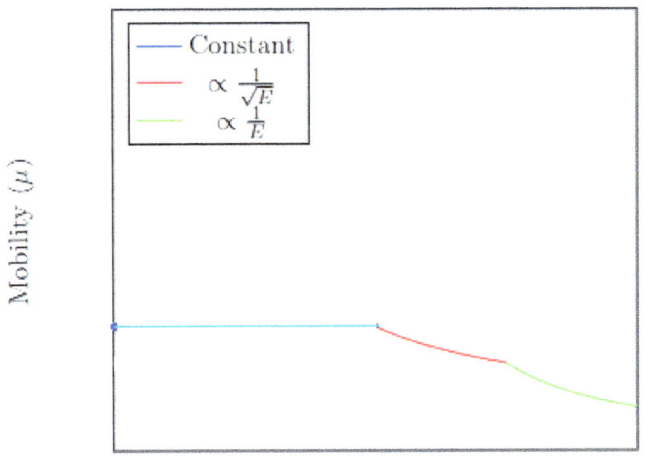

Figure 3 Variation of Mobility with Electric Field

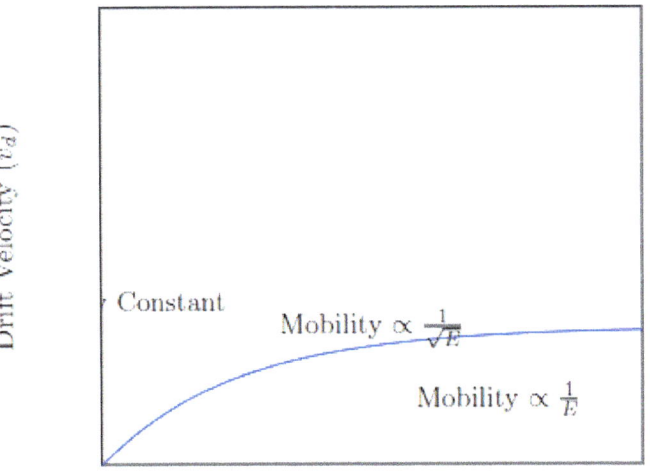

Figure 4 Variation of Drift Velocity with Electric Field

d. **Current (I):**

- Represents the flow of electric charge through a conductor.
- **Relationship to other factors:** $I = n * A * q * v_d$, where:
    - n: Charge carrier concentration (m^{-3}).
    - A: Cross-sectional area of the conductor (m^2).
    - q: Charge of the carrier (C).

Combined Effects:

- Higher mean free path (λ) typically leads to higher mobility (μ) due to fewer collisions, ultimately influencing current (I) through the relationship with drift velocity (v_d).
- A stronger electric field (E) directly increases drift velocity (v_d), thereby increasing current (I) even if mobility remains constant[14,16].
- Temperature (T) generally has a complex relationship with current (I) depending on the material and specific mechanisms involved. While higher temperature can sometimes increase carrier concentration (n), it often decreases mobility (μ), potentially leading to a decrease in current.

1.6.1 Drift-Diffusion Model

The drift-diffusion model combines the effects of carrier diffusion and carrier drift due to an electric field. It's described by the continuity equations, which include equations for carrier continuity and current density:

$$\partial n / \partial t + \nabla \cdot Jn = G - R$$

$$\partial p / \partial t + \nabla \cdot Jpp = G - R$$

where Jn and Jp are the electron and hole current densities, G is the generation rate of electron-hole pairs, and R is the recombination rate[18,19,20].

Drift Current Density (J): Drift current density represents the flow of charge carriers in response to an electric field.

For electrons in an n-type semiconductor, the drift current density (Jn) can be expressed as: $J_n = q \cdot n \cdot \mu_n \cdot E$

For holes in a p-type semiconductor, the drift current density (J_p) can be expressed as:

$$J_p = q \cdot p \cdot \mu_p \cdot E$$

where q is the elementary charge, n and p are the electron and hole concentrations respectively, μ_n and μ_p are the electron and hole mobilities respectively, and E is the electric field.

Diffusion Current Density (J_{diff}): Diffusion current density arises due to the concentration gradient of charge carriers[4,22,23].

For electrons in an n-type semiconductor, the diffusion current density ($J_{diff\,n}$) can be expressed $J_{diff\,n} = q \cdot D_n \cdot \frac{dn}{dx}$

For holes in a p-type semiconductor, the diffusion current density ($J_{diff\,p}$) can be expressed as:

$$J_{diff\,p} = q \cdot D_p \cdot \frac{dp}{dx}$$

Total Current Density (J_{total}): Drift current density +Diffusion current density

1.7 Semiconductor in Equilibrium vs. Non-Equilibrium

Equilibrium:

- A **semiconductor in equilibrium** refers to a state where the system is **stable and not undergoing any net change**[21]. This implies:

- No external forces are applied (e.g., electric field, light).
- **Charge carrier concentrations** (electrons and holes) are constant throughout the material.
- **Fermi level** is at a well-defined position based on doping and temperature.
- **No net current flow** occurs.

Some of the equations described in this state:

Poisson's Equation: Describes the relationship between electric field (E), charge density (ρ), and permittivity[25] (ε):

$$\nabla \cdot E = \frac{\rho}{\varepsilon}$$

Charge Neutrality: Ensures equal positive and negative charge densities:

$$p + n + N_d - N_a = 0$$

Mass Action Law: Relates equilibrium carrier concentrations to the intrinsic carrier concentration (n_i) and Fermi level[21] (μ):

$$np = n_i^2 = \exp\left(-\frac{\Delta E_g}{kT}\right)$$

where:

- ΔE_g: Bandgap energy
- k: Boltzmann constant
- T: Temperature

Non-Equilibrium:

- A **semiconductor in non-equilibrium** refers to a state where the system is **disturbed** from its equilibrium condition[15]. This can be caused by:
    - **Applying an external force** (e.g., electric field, light).

- **Introducing additional carriers** (e.g., through injection or generation).
- **Changing temperature** (altering thermal generation and recombination rates).
- **Presence of non-uniformities** (e.g., defects, gradients).

Non-equilibrium conditions lead to various phenomena in semiconductors, including:

- **Current flow:** When an electric field is applied, electrons and holes drift in opposite directions, resulting in current[16].
- **Carrier generation and recombination:** Light or external stimuli can excite electrons from the valence band to the conduction band, creating electron-hole pairs. These pairs can recombine, releasing energy[17] (e.g., light emission).
- **Diffusion:** Carriers can move due to concentration gradients, attempting to reach equilibrium.

Continuity Equation: Describes the change in carrier concentration over time due to generation[18], recombination, and net current flow:

$$\frac{\partial n}{\partial t} = -\nabla \cdot J_n + G_n - R_n$$

where:

- J_n: Electron current density
- G_n: Net generation rate of electrons
- R_n: Net recombination rate of electrons

Drift-Diffusion Equation: Relates current density to electric field, mobility, and carrier concentration:

$$J_n = qn\mu_n E$$

where:

- q: Electron charge
- μn: Electron mobility
- Dn: Electron diffusion coefficient

Generation-Recombination Equations: Describe the rates of carrier generation and recombination depending on various mechanisms and material properties. These equations can be complex and vary depending on the specific context[19].

$$G_n = \alpha_n I_p$$

$$R_n = \beta_n n_p$$

where:

- α_n: Recombination coefficient for electrons
- β_n: Recombination coefficient for electron-hole pairs
- I_p: Hole current density

Note: specific generation-recombination equations can vary depending on the model and mechanisms involved

1.8 Resistivity of Semiconductors

Resistivity (ρ), measured in ohmmeters (Ωm), quantifies[20] a material's resistance to the flow of electric current. Semiconductors exhibit **intermediate resistivity values** compared to metals and insulators, making them unique and valuable for various electronic applications.

Understanding Semiconductor Resistivity:

- **Intrinsic Semiconductors:** In pure, undoped semiconductors, resistivity is relatively high due to:
    - **Limited number of free charge carriers:** Thermal excitation generates some electron-hole pairs, but the concentration remains relatively low compared to metals[11,12,13,14].

- **Scattering mechanisms:** Collisions with phonons, impurities, and defects hinder carrier movement further increasing resistivity[15].
- **Doping:** Introducing dopant atoms significantly impacts resistivity:
    - **n-type doping:** Donor atoms introduce additional free electrons, **decreasing resistivity**.
    - **p-type doping:** Acceptor atoms create holes, also **decreasing resistivity** compared to intrinsic state.
- **Temperature Dependence:** Resistivity of semiconductors generally **decreases with increasing temperature** due to:
    - **Thermal generation:** More electron-hole pairs are generated, increasing carrier concentration and conductivity.
    - **Reduced scattering:** Thermal vibrations can slightly reduce the impact of scattering mechanisms[16].

Equations:
- **Resistivity and Conductivity:** The relationship between resistivity and conductivity (σ) is:

$$\rho = 1 / \sigma$$

- **Conductivity and Carrier Concentration:** Conductivity depends on carrier concentration (n for electrons, p for holes), mobility (μ), and charge (q):

$$\sigma = q\,(n\,\mu_n + p\,\mu_p)$$

Differences from Metals and Insulators:
- **Metals:** Possess **very low resistivity** due to:

- - **Abundant free electrons:** Metallic bonding results in a "sea" of mobile electrons.
  - **Weak scattering:** Relatively low collision frequency with lattice ions.
- **Insulators:** Exhibit **extremely high resistivity** because:
  - **Minimal free carriers:** Very few electrons or holes available for conduction.
  - **Strong scattering:** Tightly bound electrons and large bandgap hinder carrier movement.

Resistivity Values: Metals: 10^-8 - 10^-6 Ωm, Semiconductors: 10^-4 - 10^8 Ωm, Insulators: 10^8 - 10^23 Ωm

Semiconductor resistivity is **tunable** through doping and temperature, making them versatile materials for various applications like transistors, diodes, and integrated circuits. Their intermediate resistivity allows for controlled current flow, enabling them to function as both conductors and insulators under specific conditions. Understanding the factors influencing resistivity is crucial for designing and optimizing semiconductor devices.

1.9 Diffusion Length and Equation

Diffusion length (L), denoted[17] by the symbol "L", is a parameter used in semiconductor physics to characterize the average distance that a minority carrier (electron in a p-type or hole in an n-type) can travel before recombining with a majority carrier. It quantifies the **effectiveness of diffusion** as a transport mechanism for minority carriers.

The diffusion length is related to the **diffusion coefficient (D)** and the **minority carrier lifetime (τ)** through the following equation:

$$L = (D_p \tau_p)^{0.5}$$

$$L = (D_n \tau_n)^{0.5}$$

where:

- **L:** Diffusion length (m)
- **D:** Diffusion coefficient (m²/s)
  - **Diffusion coefficient (D):** Describes the rate at which minority carriers diffuse due to random thermal motion. Higher D implies faster diffusion and potentially longer diffusion lengths.
- **τ:** Minority carrier lifetime (s)
  - **Minority carrier lifetime (τ):** Represents the average time a minority carrier exists before recombining with a majority carrier. Longer lifetimes allow carriers to travel further before recombination, leading to a larger diffusion length.

Understanding Diffusion Length:

- A **larger diffusion length** indicates that minority carriers can travel farther before recombining, contributing more effectively to current flow within the semiconductor device.
- Conversely[18], a **smaller diffusion length** signifies limited travel distance due to frequent recombination, potentially affecting device performance.

Factors Affecting Diffusion Length:

- **Material properties:** D and τ depend on the specific semiconductor material, its doping level, and defect density.
- **Temperature:** Generally, both D and τ increase with temperature, leading to an increase in diffusion length.
- **Electric field:** Electric fields can influence carrier movement and recombination rates, potentially affecting the effective diffusion length.

Applications:

- Analyzing the performance[8,9,10] of various semiconductor devices, such as solar cells, LEDs, and transistors, often involves considering the diffusion length of minority carriers.

- Understanding diffusion length is crucial for optimizing device design and operation by controlling factors like doping and material properties.

1.10 The Einstein Relation- Are Diffusion Coefficient and mobility independent?

No, the diffusion coefficient (D) and mobility (µ) are not entirely independent. They are related through the **Einstein relation**, which describes the connection between the microscopic random motion of particles and their macroscopic transport properties[18].

Einstein Equation:

The Einstein relation for non-degenerate semiconductors at constant temperature (T) is:

$$D = \mu kT / q$$

where:

- **D:** Diffusion coefficient (m^2/s)
- **µ:** Mobility ($m^2/V \cdot s$)
- **k:** Boltzmann constant (J/K)
- **T:** Temperature (K)
- **q:** Charge of the carrier (e for electrons, -e for holes)

Explanation:

- The equation suggests that the **diffusion coefficient is proportional to the mobility and the temperature.**

- **Mobility** reflects the ease with which a carrier can move under the influence of an electric field, while **diffusion** describes[19] the random movement of carriers due to thermal energy.
- The **Einstein relation** establishes a link between these two seemingly different processes, highlighting the underlying connection between microscopic and macroscopic behavior.

Limitations:

- The Einstein relation is strictly valid for **non-degenerate semiconductors** and **low electric fields**.
- In **degenerate semiconductors** where carrier statistics deviate from classical behavior, the relationship between D and μ becomes more complex.
- **High electric fields** can also invalidate the simple form of the Einstein relation due to non-linear effects in carrier transport.

Implications:

- While not strictly independent, D and μ provide different perspectives on carrier transport:
  - **Mobility** focuses on the directed movement under an electric field.
  - **Diffusion coefficient** emphasizes the random, thermally driven motion.
- Understanding both parameters is crucial for analyzing various phenomena in semiconductors, such as carrier transport[20], device performance, and material characterization.

1.11 Band Bending

In Solid-state physics, **band bending**[21] refers to the process in which the electronic band structure in a material curves up or down near a junction or interface. It does not involve any physical bending of the material itself.

Why it happens:

- When two different materials with dissimilar electrochemical potentials for their free charge carriers are brought into contact, charge carriers tend to flow from the material with higher potential to the one with lower potential until an equilibrium state is reached.

- This flow of charge creates an **electric field** at the interface, which in turn affects the potential energy of the charge carriers in the nearby regions of each material.

Consequences of band bending:

- The **curvature of the bands** reflects the change in potential energy experienced by the charge carriers.

- In **n-type semiconductors**, the conduction band typically bends **upward**, making it more difficult for electrons to move towards the interface.

- Conversely, in **p-type semiconductors**, the valence band usually bends **downward**, hindering the movement of holes towards the interface.

Impact on various phenomena:

- Band bending plays a crucial role in various semiconductor devices, including:

    - **Diodes:** The direction and magnitude of band bending determine the current flow characteristics of a diode.

    - **Transistors:** Band bending in different regions of a transistor controls the flow of charge carriers and enables switching and amplification functions.

    - **Solar cells:** The band bending at the p-n junction is essential for efficient light absorption and charge separation in solar cells.

Additional points:

- Band bending is not limited to semiconductors and can also occur in other materials like insulators and metals at interfaces.

- The specific degree and direction of band bending depends on the properties of the materials in contact, the doping levels, and the applied external factors like electric fields.

1.12 Electrical Analogy of Scattering in Semiconductors

Imagine a semiconductor as a circuit:

Electrons (or holes) are like charged particles flowing through the circuit. Scattering centers (impurities, defects, phonons[21]) act like obstacles or resistors in the circuit.

When electrons encounter scattering centers:

- Their movement is impeded, similar to how resistors hinder current flow.

- They may change direction or lose energy, analogous to voltage drops across resistors.

Types of Scattering and their Analogies:

1. Impurity Scattering:

    o Analogy: Like encountering a fixed resistor in the circuit, representing the presence of dopant atoms.

    o Effect: Reduces electron mobility and conductivity.

2. Defect Scattering:

    o Analogy: Like encountering a faulty or damaged component in the circuit, representing imperfections in the crystal structure.

    o Effect: Can significantly hinder electron movement and increase resistivity.

3. Phonon Scattering:

    - Analogy: Imagine encountering random fluctuations in the circuit's resistance due to thermal vibrations, representing lattice vibrations (phonons).
    - Effect: Increases with temperature as phonon activity intensifies, impacting mobility and conductivity[25].

Consequences of Scattering:

- Reduced mobility: Scattering events impede the average velocity of electrons, leading to lower mobility.
- Increased resistivity: Lower mobility[11] translates to higher resistance to current flow.
- Material dependence: The type and density of scattering centers significantly influence the electrical properties of different semiconductors.

1.13 Optical Processes in Semiconductors: Absorption and Emission

Semiconductors exhibit unique optical properties due to their band structure and the interaction between light and electrons. Two fundamental processes govern this interaction:

1. Absorption:

- **Mechanism:** When a photon with sufficient energy strikes a semiconductor, it can be absorbed by an electron in the valence band.
- **Energy transfer:** The absorbed photon's energy excites the electron, promoting it to a higher energy level in the conduction band.
- **Result:** This creates an electron-hole pair, where the "hole" represents the vacancy left behind in the valence band.

Factors affecting absorption:

- **Photon energy:** The photon's energy must be equal to or greater than the bandgap energy (energy difference between valence and conduction bands) for absorption to occur.

- **Material properties:** Bandgap energy varies depending on the semiconductor material, influencing the wavelengths of light it can absorb.

- **Doping level:** Doping can slightly alter the bandgap, affecting the absorption spectrum.

2. Emission:

- **Mechanism:** There are several ways for an excited electron to return to its ground state and emit light:

    - **Spontaneous emission:** The excited electron relaxes back to the valence band, releasing energy as a photon.

    - **Stimulated emission:** An external photon stimulates the transition, causing the emission of another photon with the same energy and phase. This principle forms the basis of lasers.

    - **Non-radiative recombination:** The electron loses energy through collisions with phonons (lattice vibrations) without emitting light.

Factors affecting emission:

- **Recombination[16] process:** The dominant emission mechanism (spontaneous vs. stimulated) depends on various factors like material properties, doping levels, and external stimuli.

- **Emission wavelength:** The emitted photon's energy corresponds to the energy difference between the initial and final states of the electron, determining the emitted light's wavelength (color).

Applications of these processes:

Photodetectors: Utilize absorption to convert light into electrical signals, essential for solar cells, photodiodes, and optical communication devices[13].

Light-emitting-diodes-(LEDs): Employ, electroluminescence, where electrical energy is converted into light through stimulated emission.

Lasers: Rely on stimulated emission to achieve highly coherent and directional light output.

1.14 Ambipolar Transport in Semiconductors

Ambipolar transport refers to the simultaneous movement of both **electrons** and **holes** in opposite directions within a semiconductor material. This phenomenon arises due to the presence of **excess carriers** that are not part of the material's intrinsic equilibrium[15].

Understanding the Concept:

- In **intrinsic semiconductors**, the number of mobile charge carriers (electrons and holes) is relatively low due to thermal generation. Therefore, ambipolar transport plays a minimal role.

- However, when a semiconductor is **doped** or subjected to external stimuli like **illumination**, the concentration of mobile carriers increases significantly. This creates **excess carriers** beyond the thermal equilibrium levels.

- These excess carriers, driven by their concentration gradients and electric fields, can contribute to both electron and hole flow, leading to ambipolar transport[17].

Factors Influencing Ambipolar Transport:

- **Carrier concentration:** Higher concentrations of excess carriers enhance ambipolar transport by providing more mobile charges for both types of carriers.

- **Material properties:** Semiconductor's band structure, mobility of electrons and holes, and diffusion coefficients all influence the relative contributions of each carrier type to the overall transport.

- **Electric field:** An applied electric field[11] can influence the direction and magnitude of both electron and hole movement, impacting the overall ambipolar transport characteristics.

Applications of Ambipolar Transport:

- **Transistors:** In certain transistor operating regions, ambipolar transport can contribute to leakage currents and affect device performance. Understanding and controlling this phenomenon is crucial for optimal transistor design.

- **Organic semiconductors:** Ambipolar transport is observed in some organic semiconductors, making them potentially suitable for applications like organic light-emitting diodes (OLEDs) and organic field-effect transistors (OFETs)[10].

- **Sensor devices:** By controlling ambipolar[3] transport through material selection and device design, researchers are exploring possibilities for developing novel sensors sensitive to various stimuli.

It's important to note that:

- Ambipolar transport may not always be desirable, as it can lead to unwanted leakage currents and impact device performance[5].

- In many semiconductor devices, strategies are employed to minimize ambipolar transport and favor the dominant flow of one type of carrier (electrons or holes) for intended functionality[6].

1.15 Hall Effect: Explanation with Equations and Diagrams

The **Hall effect** describes the emergence of a voltage **perpendicular** to both an applied current and a magnetic field in a conductor. This phenomenon arises due to the **Lorentz force** acting on moving charge carriers within the conductor[11,12,13,14].

Explanation:

1. **Current flow:** When a current (I) is passed through a conductor (rectangular in this case), electrons (or holes) drift in a specific direction, as shown in the diagram.

2. **Magnetic field:** When a magnetic field (B) is applied perpendicular to the current flow and the conductor, it exerts a **Lorentz force** on the moving charge carriers.

3. **Lorentz force:** The Lorentz force (F) is given by:

$$F = q (v \times B)$$

where:

- q is the charge of the carrier (e for electrons, -e for holes)
- v is the drift velocity of the carriers
- x denotes the cross product

4. **Hall voltage:** Due to the Lorentz force, the moving charges experience a **lateral deflection** towards one side of the conductor. This accumulation of charges on one side and depletion on the other side creates a **potential difference** across the width (d) of the conductor, known as the **Hall voltage (V_H)**.

Equilibrium: The Hall voltage builds up until the electric field it creates balances the Lorentz force, establishing equilibrium[18].

Equation for Hall Voltage:

The Hall voltage can be calculated using the following equation:

$$V_H = (R_H * I * B) / d$$

where:

- R_H is the **Hall coefficient**, a material property reflecting the type and concentration of charge carriers.

$$R_H = \text{Mobility/ Conductivity}$$

- I is the current flowing through the conductor.
- B is the magnetic field strength.
- d is the width of the conductor.

Chapter-2

P-N Diode

2.1 Introduction to P-N Diode

A PN junction diode is one of the simplest semiconductor devices, consisting of a P-type semiconductor material and an N-type semiconductor material joined together. Here's a brief overview of its basic structure and working principle:

1. **Basic Structure:**

    - The PN diode is typically formed by doping a semiconductor material such as silicon or germanium with impurities to create regions of P-type and N-type material[19].

    - The P-type region is doped with a trivalent impurity, resulting in an excess of positive "holes" as majority carriers.

    - The N-type region is doped with a pentavalent impurity, creating an excess of negative charge carriers, usually electrons, as majority carriers.

    - When the P-type and N-type materials are brought into contact, a junction is formed between them known as the PN junction[20].

2. **Working Principle:**

    - At equilibrium, a built-in potential barrier is established across the PN junction due to the diffusion of charge carriers.

    - In the P-type region, the concentration of holes is higher than that of electrons, creating a region with a net positive charge[21].

- Conversely, in the N-type region, the concentration of electrons is higher than that of holes, resulting in a net negative charge.

- The built-in potential barrier prevents further diffusion of majority carriers across the junction, establishing an electric field that opposes the flow of charge.

- When a forward bias voltage is applied across the PN junction (positive to the P-side, negative to the N-side), it reduces the width of the depletion region, allowing majority carriers to overcome the barrier and flow across the junction.

- This results in a forward current flow through the diode, with electrons moving from the N-side to the P-side and holes moving from the P-side to the N-side.

- Conversely, when a reverse bias voltage is applied (negative to the P-side, positive to the N-side), it widens the depletion region, increasing the barrier height and preventing significant current flow.

- However, a small reverse current known as the leakage current may still occur due to minority carrier diffusion and thermally generated carriers.

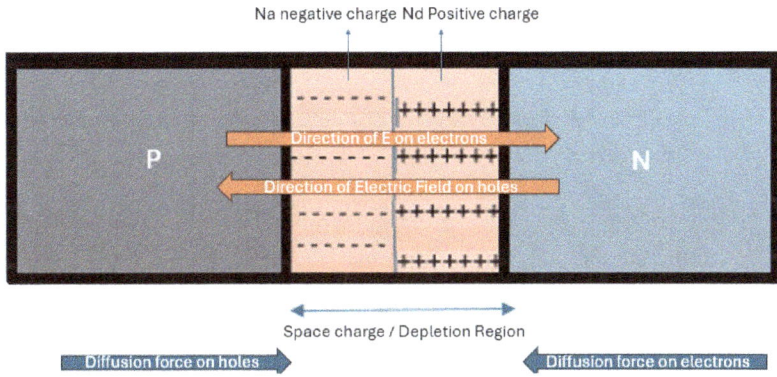

Figure 5 Space charge region in p-n junction Diode

In thermal equilibrium, a p-n diode establishes a steady-state condition where the flow of majority carriers (holes in the p-region and electrons in the n-region) across the junction balances out the flow of minority carriers (electrons in the p-region and holes in the n-region) due to diffusion. This equilibrium is maintained by the built-in potential barrier at the junction.

Additionally, the electric field established by the built-in potential prevents further diffusion of majority carriers across the junction. The built-in potential Vbi is determined by the difference in Fermi levels between the p and n regions and is given by[11]:

$$n_p(x) \cdot p_n(x) = n_i^2$$

Where:

- $n_p(x)$ is the electron concentration in the p-region,
- $p_n(x)$ is the hole concentration in the n-region,
- ni is the intrinsic carrier concentration.

$$V_{bi} = \frac{kT}{q} \ln\left(\frac{N_d \cdot N_a}{n_i^2}\right)$$

where:

- k is Boltzmann's constant,
- T is the temperature in Kelvin,
- q is the elementary charge,
- Nd and Na are the doping concentrations of the n and p regions, respectively.

2.1.1 Diode Current Equation (Shockley Diode Equation)

The diode current, I_D, in a forward biased PN junction diode can be described by the Shockley diode equation[11]:

$$I_D = I_0 \left(e^{\frac{V_D}{nV_T}} - 1\right)$$

where:

- I_0 is the reverse saturation current,
- V_D is the voltage across the diode,
- V_T is the thermal voltage, approximately kT/q at room temperature (k is the Boltzmann constant, T is the temperature in Kelvin, and q is the elementary charge),
- n is the ideality factor, typically around 1 for silicon diodes.

2.1.2 Diode Voltage Equation (Ideal Diode Equation)

In reverse bias, the diode equation simplifies to:

$$I_D = I_0 \times e^{\frac{V_D}{nV_T}}$$

Since V_D is negative in reverse bias, $e^{\frac{V_D}{nV_T}}$ approaches zero exponentially, resulting in a very small reverse current[12].

2.1.3 Diode Resistance

The dynamic resistance of a diode, rd, can be approximated by the inverse of the slope of the diode I-V curve at a given operating point[13]:

$$r_d = \frac{\Delta V_D}{\Delta I_D}$$

This resistance is typically very small in forward bias and very large in reverse bias.

2.1.4 Diode Capacitance

The depletion capacitance of the diode, C_D, can be calculated as the ratio of the charge stored in the depletion region to the change in voltage[14]:

$$C_D = \frac{dQ}{dV}$$

where dQ is the change in charge and dV is the change in voltage.

2.1.5 Space Charge Width

The space charge width refers to the region within a semiconductor device where the majority carriers (electrons or holes) and minority carriers (holes or electrons, respectively) create a net electric field due to their spatial distribution and resulting charge imbalance[15].

Equation: The space charge width (W) can be mathematically described using the following equation:

$$W = \sqrt{\frac{2\epsilon\epsilon_0}{qN_d}(V_{bi} - V)}$$

$$W = \sqrt{\frac{2V_{bi}\epsilon\epsilon_0}{q}\left(\frac{N_a + N_d}{N_a N_d}\right)}$$

Where:

- W = space charge width
- ϵ = permittivity of the semiconductor material
- ϵ_0 = permittivity of free space
- q = elementary charge
- N_d = doping concentration of the semiconductor material (donor: positive Charge)
- N_a = doping concentration of the semiconductor material (acceptor: negative Charge)
- V_{bi} = built-in potential (also known as barrier potential or junction potential)
- V = applied voltage across the semiconductor device

Explanation:

1. **Doping Concentration (Nd):** This term represents the concentration of dopant atoms introduced into the semiconductor

material. It determines the density of charge carriers within the material[16].

2. **Built-in Potential (*Vbi*):** This is the potential difference established across the semiconductor junction due to the difference in the work functions of the two materials forming the junction. It arises from the redistribution of charge carriers during the formation of the junction.

3. **Applied Voltage (*V*):** The external voltage applied to the semiconductor device. This voltage alters the width of the space charge region by modifying the electric field within the device[17].

4. **Permittivity (*ϵ*):** The permittivity of the semiconductor material, which describes its ability to store electrical energy in an electric field. It depends on the material's composition and temperature.

5. **Elementary Charge (*q*):** The fundamental charge of an electron.

The equation illustrates that the space charge width is influenced by both the intrinsic properties of the semiconductor material (such as doping concentration and permittivity) and the external factors (such as applied voltage). As the applied voltage changes, the width of the space charge region adjusts accordingly, impacting the device's overall behavior, such as its conductivity and capacitance[20].

2.2 Reverse Bias with P-N Diode

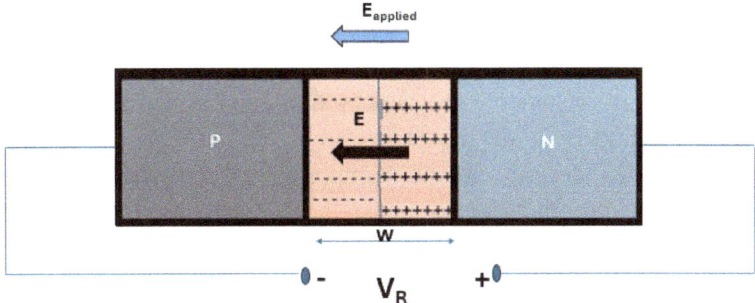

Figure 6 Reverse Bias in p-n junction Diode (n connected positive and p connected negative)

$$V_{total} = V_{bi} + V$$

- V_{total} = Total potential barrier for p-n junction
- V_{bi} = built-in potential (also known as barrier potential or junction potential)
- V_R = applied Reverse Bias voltage across the semiconductor device

When a reverse bias voltage (Vr) is applied across a semiconductor diode, it increases the potential difference across the depletion region. This increase in potential difference causes the width of the depletion region to expand. The built-in potential (Vbi) represents the potential difference established across the depletion region in the absence of an external bias[20].

In reverse bias, the electric field (E) within the depletion region is directly proportional to the change in voltage (dV) across the region and inversely proportional to the change in distance (dx) across the depletion region. Mathematically, this can be represented as:

$$E = dx/dV$$

In terms of reverse bias voltage (Vr) and built-in potential (Vbi), the change in voltage (dV) across the depletion region can be expressed as the difference between the applied reverse bias voltage (Vr) and the built-in potential (Vbi). Therefore, dV=Vr−Vbi.

The change in distance (dx) across the depletion region is not explicitly included in the equation, as it depends on the specific geometry of the device and the semiconductor material properties. However, it's important to note that in reverse bias, the depletion region typically widens due to the applied voltage, resulting in an increase in dx[26].

Combining these relationships, the equation for the electric field (E) in terms of reverse bias voltage (Vr) and built-in potential (Vbi) can be written as:

$$E = \frac{Vr + Vbi}{dx}$$

Using the equations above and replacing to get values in terms of W

$$E = \frac{-2(Vbi + Vr)}{W}$$

2.2.1 Difference between Junction Capacitance and Diode capacitance

Junction capacitance and diode capacitance are related concepts that describe the capacitance associated with a semiconductor diode, particularly across its p-n junction. However, they represent **different aspects** of the diode's behavior[24].

Junction Capacitance: Junction capacitance refers to the capacitance that exists across the depletion region of a semiconductor diode. In a p-n junction diode, when a voltage is applied in the reverse bias direction, the depletion region widens, creating a space charge region with a built-in electric field.

This region acts as a dielectric between the p-type and n-type semiconductor materials, effectively forming a capacitor. Junction capacitance arises due to the separation of charge carriers within the depletion region, resulting in a capacitance that opposes changes in voltage across the diode[5,6].

Mathematically, the junction capacitance (Cj) is inversely proportional to the width of the depletion region (W) and directly proportional to the permittivity of the semiconductor material (ϵ):

$$C_j = \frac{\epsilon \epsilon_0}{W}$$

The width of the depletion region (W) can be expressed in terms of the built-in voltage and reverse bias voltage using the following relationship:

$$W = \sqrt{\frac{q}{2\epsilon\epsilon_0}\left(\frac{1}{N_a}+\frac{1}{N_d}\right)(V_{bi}-V_r)}$$

Substituting the expression for W into the equation for Cj, we obtain:

$$C_j = \frac{\epsilon\epsilon_0}{\sqrt{\frac{q}{2\epsilon\epsilon_0}\left(\frac{1}{N_a}+\frac{1}{N_d}\right)(V_{bi}-V_r)}}$$

Simplifying the expression, we get:

$$C_j = \sqrt{\frac{q}{2}\epsilon\left(\frac{1}{N_a}+\frac{1}{N_d}\right)\frac{1}{V_{bi}-V_r}}$$

Diode Capacitance: Diode capacitance, on the other hand, is a more general term that encompasses various capacitance effects associated with a semiconductor diode[7,8]. It includes not only the junction capacitance but also other capacitance components, such as the diffusion capacitance and the transition capacitance[11,12].

1. **Junction Capacitance:** As described above, this is the capacitance associated with the depletion region of the diode.

2. **Diffusion Capacitance:** This capacitance arises due to the storage and release of charge carriers within the semiconductor's neutral regions as the diode transitions between conducting and non-conducting states[13].

$$C_{diff} = \frac{Q}{V}$$

Where:

- C_{diff} is the diffusion capacitance.
- Q is the charge stored or released.

- V is the voltage across the diode.

3. **Transition Capacitance:** Also known as the depletion capacitance, this component arises due to the change in depletion region width with changes in applied voltage (especially during signal transitions)[14].

$$C_{trans} = \frac{dQ}{dV}$$

Where:

- C_{trans} is the transition capacitance.

- dQ/dV represents the rate of change of charge with respect to voltage.

2.3 Forward Bias with P-N Diode

"Forward bias" refers to the application of a voltage across the diode such that the positive terminal of the voltage source is connected to the P-type semiconductor and the negative terminal is connected to the N-type semiconductor.

When a PN junction diode is forward biased, it allows current to flow easily through it, as opposed to reverse bias where it blocks the flow of current. This is because the forward bias reduces the width of the depletion region at the junction, making it easier for charge carriers (electrons and holes) to cross the junction and contribute to the flow of current.

Under forward bias:

1. Electrons from the N-type region are attracted towards the positive terminal of the voltage source.

2. Holes from the P-type region are attracted towards the negative terminal of the voltage source.

3. The electrons and holes move towards the junction and recombine, allowing current to flow through the diode.

4. Additionally, the voltage applied forward biases the diode, overcoming the potential barrier created by the built-in electric field at the junction.

The forward bias condition is crucial for the operation of diodes in electronic circuits, such as rectifiers, where they allow current to flow in one direction while blocking it in the opposite direction.

2.3.1 Excess Carrier Concentration

$$\Delta n = \frac{D_n \cdot \Delta p_0 + D_p \cdot \Delta n_0 + \delta \cdot \Delta n_0}{W^2 + 2\delta \cdot W}$$

The equation represents the excess carrier concentration, denoted by Δn, in a semiconductor device.

- Δn: Excess electron concentration, which is the additional number of electrons in the semiconductor compared to its equilibrium state[12].

- Δp_0: Equilibrium hole concentration, representing the number of holes (missing electrons) in the semiconductor under equilibrium conditions.

- Δn_0: Equilibrium electron concentration, representing the number of electrons in the semiconductor under equilibrium conditions.

- D_n: Electron diffusion coefficient, a parameter that describes how quickly electrons move in response to a concentration gradient.

- D_p: Hole diffusion coefficient, similar to D_n but for holes[22].

- δ: Defect concentration, representing the concentration of defects or impurities in the semiconductor material.

- W: Width of the depletion region, which is the region near the PN junction where there are no free charge carriers due to the presence of an electric field.

The equation describes how the excess carrier concentration (Δn) depends on various factors such as the diffusion coefficients of electrons (Dn) and holes (Dp), the equilibrium concentrations of electrons (Δn_0) and holes (Δp_0), the defect concentration (δ), and the width of the depletion region (W). It shows how these factors contribute to the generation and recombination of excess carriers within the semiconductor device[23].

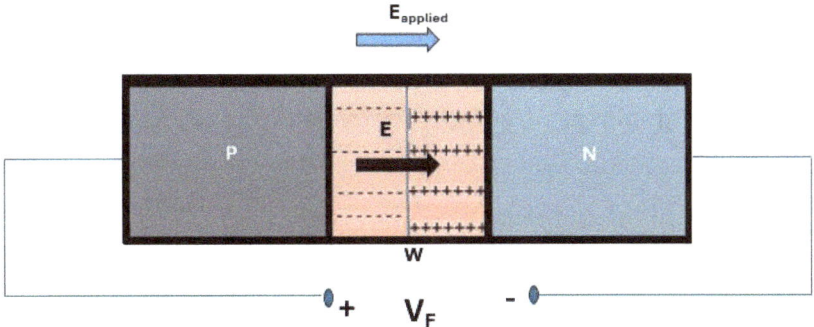

Figure 7 Forward Bias in p-n junction Diode (P connected positive and N connected negative)

From section 1.7, law of mass action, we can rewrite the excess charge concentration in P-N diode with forward Voltage V_F:

$$N = n_o \exp\left(\frac{eVF}{kT}\right)$$

$$p = p_o \exp\left(\frac{eVF}{kT}\right)$$

2.3.2   Forward Bias Resistance, capacitance, and admittance

2.3.2.1 Resistance
- Resistance is a measure of how much a material or device opposes the flow of electric current. In the case of a diode, it can be categorized into two states: forward resistance and reverse resistance.

- Forward resistance (R_f) refers to the resistance offered by a diode when it is forward biased.

- Reverse resistance (R_r) refers to the resistance offered by a diode when it is reverse biased.

- Mathematically, resistance can be expressed as: $R=IV$ Where V is the voltage across the diode and I is the current flowing through it.

- Example: If a diode has a forward voltage of 0.70 volts and a forward current of 10 mA, its forward resistance would be 70 ohms[24].

2.3.2.2 Capacitance

Capacitance is a measure of a device's ability to store electrical charge when a voltage difference exists between its terminals. In the context of a diode, capacitance arises due to the depletion region's ability to store charge.

- Capacitance is a measure of a device's ability to store electrical charge when a voltage difference exists between its terminals. In the context of a diode, capacitance arises due to the depletion region's ability to store charge[25].

- The junction capacitance (C_j) of a diode varies with the reverse bias voltage applied across it. It can be modeled using the diode junction capacitance equation[26]:

$$C_j = \frac{C_0}{\sqrt{1 + \frac{V_{reverse}}{V_{breakdown}}}}$$

- Where C_0 is the junction capacitance at zero bias, $V_{reverse}$ is the reverse bias voltage, and $V_{breakdown}$ is the breakdown voltage of the diode.

2.3.2.3 Admittance

Admittance is the reciprocal of impedance and represents a circuit's ability to allow current to flow. In the case of a diode, admittance can be characterized as dynamic due to the varying behavior of the diode with applied voltage[11].

Small signal admittance refers to the linearized conductance and susceptance of a device or circuit around a certain operating point. In electronic circuits, particularly in the field of linear circuit analysis, small signal models are used to analyze the behavior of components around their bias or operating points[7].

Small signal admittance is a measure of how a circuit responds to small changes or perturbations in voltage or current around its operating point. It is typically represented as a linearized model derived from the small signal parameters of the device or circuit.

For a two-terminal device like a diode, small signal admittance can be represented as a complex quantity, where the real part represents the conductance (measured in Siemens, S) and the imaginary part represents the susceptance (measured in Siemens per unit frequency, S/Hz).

The small signal admittance (Y) can be expressed as:

$$Y = G + jB$$

Where:

- G is the conductance (real part), which represents the linearized change in current with respect to voltage.

- B is the susceptance (imaginary part), which represents the linearized change in reactive behavior (such as capacitance or inductance) with respect to frequency.

Small signal admittance models are particularly useful in analyzing the frequency response of circuits, as they allow engineers to understand how a circuit's behavior changes with frequency around its operating point.

These models are commonly used in the design of amplifiers, filters, and other linear circuits[2].

Feature	Admittance	Small-Signal Admittance
Definition	The overall ease with which a circuit or device allows current to flow. It's the reciprocal of impedance.	A linearized approximation of admittance around a specific operating point (bias point). It describes how small changes in voltage affect small changes in current.
Scope	Considers both linear and non-linear circuit behavior.	Focuses specifically on the linear behavior of a circuit around a fixed operating point.
Mathematical Representation	$Y = G + jB$ where: Y = Admittance, G = Conductance (real part, represents power dissipation) B = Susceptance (imaginary part, represents energy storage)	$y = g + jb$ where: y = small-signal admittance, g = small-signal conductance, b = small-signal susceptance
Application	Analyzing general circuit behavior, including non-linear elements like diodes and transistors.	Analyzing circuits with non-linear elements when you're interested in their behavior in response to small AC voltage fluctuations around a DC bias point.
Significance	Provides a fundamental understanding of how a circuit responds to electrical signals.	Essential for modeling and designing AC circuits involving non-linear devices, such as amplifiers and oscillators.

Table 2 Differences between Admittance and small signal admittance with feature comparisons

2.4 I-V characteristics of a Diode

The IV characteristics of a diode describe its current-voltage relationship, which is crucial in understanding how a diode behaves in an electronic circuit. "IV" stands for current (I) versus voltage (V).

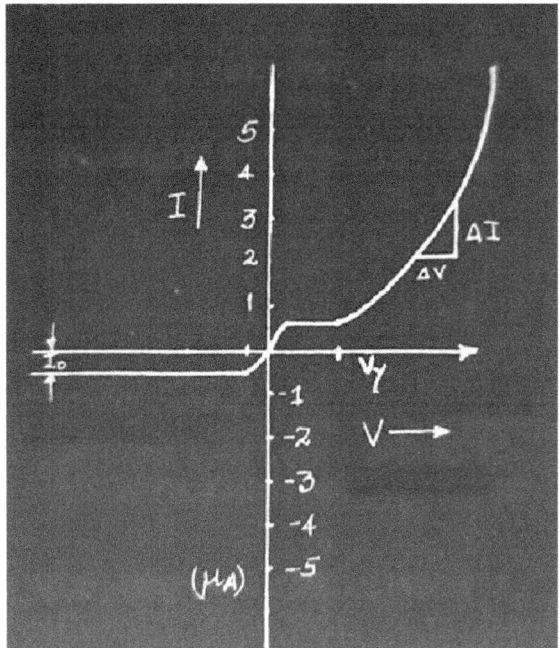

Figure 8 IV characteristics of a diode.

A diode is a two-terminal semiconductor device that allows current to flow in one direction only. It has two states: forward bias and reverse bias[26].

1. Forward Bias:

    - When a positive voltage is applied to the anode (the positive terminal) and a negative voltage to the cathode (the negative terminal), the diode is said to be forward biased.

    - In this state, the diode offers very low resistance to the flow of current. Current flows easily through the diode, and it conducts electricity.

- The voltage across the diode increases slightly with increasing current, but it remains relatively constant.

2. Reverse Bias:
    - When a negative voltage is applied to the anode and a positive voltage to the cathode, the diode is reverse biased[1,2,3].
    - In this state, the diode exhibits very high resistance to the flow of current. Only a small leakage current flows, typically in the nanoampere range.
    - The voltage across the diode increases rapidly with increasing reverse bias voltage, reaching a breakdown voltage where the diode may break down and allow significant current to flow in the reverse direction.

The IV characteristics of a diode can be graphically represented. In the forward bias region, the current increases exponentially with increasing voltage, following the diode equation:

$$I = I_0 \cdot \left(e^{\frac{V}{nV_T}} - 1\right)$$

Where:
- I is the diode current.
- I_0 is the reverse saturation current.
- V is the voltage across the diode.
- n is the ideality factor, typically between 1 and 2.
- VT is the thermal voltage, approximately 26mV at room temperature.

In the reverse bias region, the current remains very low until the breakdown voltage is reached, after which the current increases rapidly.

Semiconductor Essentials

2.5 Comparison between P-N Junction characteristics for Zero Bias, Reverse Bias, Forward Bias

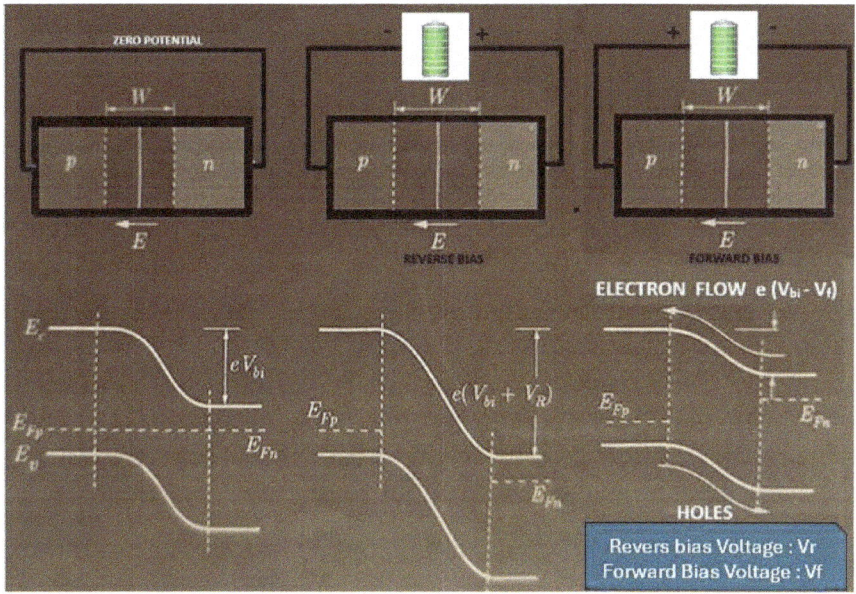

Figure 9 Comparison of P-N junction Characteristics of ZERO bias, RB (REVERSE BIAS), FB (FORWARD BIAS)

Fig shows the energy band diagram in equilibrium, followed by reverse biased p-n junction. The Fermi Level of the N region is lower than P in reverse biased P-N junction whereas it's the opposite in forward Bias. Also, Width of junction is broader in Reverse Bias[4].

2.6 Junction Breakdown

Junction breakdown is a phenomenon that occurs in **p-n junctions**, which are the basic building blocks of various electronic devices like diodes and transistors. It happens when a **reverse bias voltage** exceeding a specific threshold, called the **breakdown voltage**, is applied across the junction.

Depletion Region and Current:

- Under reverse bias, a small **reverse saturation current** flows due to thermally generated minority carriers (electrons in p-type and holes in n-type) drifting across the depletion region.

Breakdown Voltage and Mechanisms:

- As the reverse voltage increases, the electric field in the depletion region intensifies. When the voltage reaches a critical point, called the **breakdown voltage (Vbr)**, two main mechanisms can lead to a sudden increase in reverse current:

    o **Avalanche breakdown:** In **lightly doped** junctions, high electric field accelerates minority carriers, giving them enough energy to **collide with atoms** in the lattice and knock out additional electron-hole pairs. These newly created carriers are further accelerated, causing an **avalanche effect** and a sharp rise in current.

    o **Zener breakdown:** In **heavily doped** junctions, the strong electric field allows some **minority carriers** to **tunnel** through the narrow depletion region, enabling current flow even at lower voltages.

Consequences:

- If the current surge due to breakdown is not limited by circuit elements, it can **permanently damage** the pn junction due to excessive heating.

- However, in **controlled conditions**, breakdown can be utilized in specific devices like **Zener diodes** for voltage regulation or **avalanche photodiodes** for light detection[7,8,9].

2.7 Special types of PN JUNCTION DIODE

2.7.1 Tunnel Diode

A **tunnel diode**, also known as an **Esaki diode**, is a special type of semiconductor diode known for its **negative resistance** characteristic. Here's a breakdown of its key aspects:

1. Basic Functionality

- Like regular diodes, tunnel diodes are constructed from p-type and n-type semiconductors forming a **p-n junction**.

- However, unlike regular diodes that primarily conduct current in the forward bias direction, tunnel diodes exhibit interesting behavior in both **forward and reverse bias**.

2. Tunneling Effect

- The key difference lies in the **heavily doped** nature of the p-n junction in a tunnel diode. This heavy doping creates a very **narrow depletion region** between the p and n regions[10,11].

- Under certain conditions, electrons can **tunnel** through this narrow barrier due to a quantum mechanical phenomenon called **quantum tunneling**. In essence, the electrons can "pass through" the barrier even though they don't have enough classical energy to overcome it entirely.

3. Current-Voltage Characteristic

- The resulting **current-voltage (I-V) characteristic** of a tunnel diode exhibits a unique negative resistance region[13].

- At low forward bias voltages, the tunneling effect allows a **relatively high current** to flow. However, as the voltage increases further, the tunneling probability **decreases**, leading to a **decrease in current**. This creates a **negative resistance** region in the I-V curve.

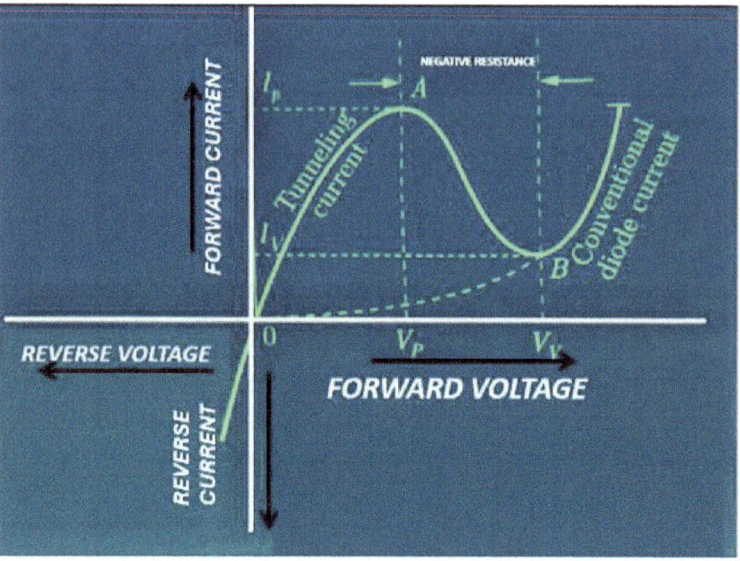

Figure 10 Current-Voltage (I-V) characteristic of a tunnel diode

4. Applications

- Due to their negative resistance characteristic and fast switching capabilities, tunnel diodes are often used in high frequency applications such as:

    o **Oscillators:** Generating high frequency signals.

    o **Mixers:** Converting signals from one frequency range to another.

    o **High-speed switches:** Switching circuits on and off rapidly.

5. Limitations

- While offering unique advantages, tunnel diodes also have limitations:

    o **Low current handling capacity:** They are not suitable for high-current applications.

- o **Temperature sensitivity:** Their performance can vary significantly with temperature changes.
- o **Complex fabrication:** Their manufacturing process is more complex compared to regular diodes.

6. *Tunnel Diode Equivalent Circuit*

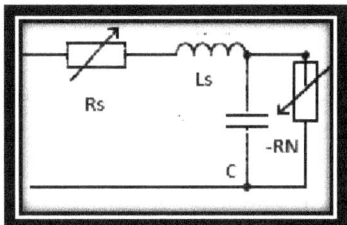

Figure 11 Tunnel diode equivalent circuit representation.

Fig above shows the equivalent circuit for Tunnel diode. 4 elements can be defined as:

1) Series Resistance (Rs) is the resistance due to leads, contacts, and semiconductor material.
2) Series Inductance (Ls) is the inductance due to lead lengths.
3) Junction Capacitance (C) is the diffusion capacitance and applied voltage.
4) Negative Resistance (-Rn) is the resistance offered by Tunnel Diode between Peak Point A and Valley point B (in I-V characteristics of Tunnel Diode)[12]

2.7.2 PIN Diode

A PIN diode is a special type of semiconductor diode with a wide, undoped **intrinsic semiconductor** region (I) sandwiched between a **p-type** semiconductor (P) and an **n-type** semiconductor (N).

This intrinsic region gives the PIN diode unique characteristics suitable for various applications.

1. Key Characteristics
- **Low Forward Resistance:** The intrinsic region offers lower resistance when the diode is forward-biased, allowing larger currents to flow for a given voltage.
- **High Reverse Breakdown Voltage:** The intrinsic region widens the depletion layer, allowing the diode to withstand higher reverse voltages before breakdown.
- **Variable Resistance:** Under forward bias, the conductivity of the intrinsic region can be controlled by the amount of current, making the PIN diode act like a variable resistor.
- **Fast Switching:** The intrinsic region reduces the diode's capacitance, allowing for very fast switching between conducting and non-conducting states[19].

2. Applications
- **RF Switches:** PIN diodes are ideal for radio frequency (RF) switching due to their fast-switching speeds, low capacitance, and ability to handle high power. They're used in circuits to switch RF signals between different paths.
- **RF/Microwave Attenuators:** The variable resistance of PIN diodes makes them useful for attenuating (reducing the amplitude of) RF and microwave signals[21].
- **Photodetectors:** PIN diodes work as good photodetectors (converting light into electrical signals). The wide intrinsic region offers a larger volume for absorbing photons, improving their light detection sensitivity[25].
- **High-Voltage Rectifiers:** Because they withstand high reverse voltages, PIN diodes can be used for rectifying high-voltage AC signals.

3. Advantages of PIN Diodes
- Ability to handle high power levels.
- Excellent RF performance
- Fast switching times

- Low distortion at high frequencies

2.7.3 Varactor Diode

A varactor diode (also known as a varicap, tuning diode, or variable capacitance diode) is a special type of semiconductor diode designed to function primarily in reverse bias.

Its key feature is that its internal junction capacitance varies depending on the reverse voltage applied across the diode.

1. How Does It Work?
1. **Depletion Region:** When a p-n junction diode is reverse-biased, a depletion region forms where there are few free charge carriers. This depletion region acts like an insulator between the conductive p and n regions, essentially forming a capacitor[9,10,11].
2. **Voltage Control:** The width of the depletion region, and therefore the capacitance of the diode, is directly affected by the applied reverse bias voltage.
    - **Higher Reverse Voltage:** Widens the depletion region, decreasing capacitance.
    - **Lower Reverse Voltage:** Narrows the depletion region, increasing capacitance.

2. Characteristics
- **Variable Capacitance:** The primary characteristic of a varactor diode, allowing it to act as a voltage-controlled capacitor.
- **Operation in Reverse Bias:** Varactor diodes are always designed to operate in reverse bias conditions to exploit this capacitance variation.
- **Non-Linear Relationship:** The relationship between capacitance and voltage is not linear[5,6,8].

3. *Typical Applications*

- **Tuning Circuits:** Used in radio frequency (RF) circuits for tuning purposes. Adjusting the reverse voltage applied to the varactor changes its capacitance, which in turn alters the resonant frequency of the circuit. Applications include radios, televisions, and other communication devices.

- **Voltage-Controlled Oscillators (VCOs):** Varactor diodes are core components in VCOs. By varying the control voltage to the diode, the oscillation frequency can be changed.

- **Frequency Multipliers:** Varactor diodes are used to multiply the frequency of an input signal due to their non-linear characteristics[14].

- **Filters:** In the design of tunable filters.

2.7.4 Schottky Diode

A Schottky diode, also known as a Schottky barrier diode or hot-carrier diode, is a type of semiconductor diode formed by the junction of a metal and a semiconductor, rather than the typical p-n junction found in standard diodes. This unique structure gives it several advantages:

1. *Key characteristics*

- **Low forward voltage drop:** Compared to standard p-n junction diodes, Schottky diodes exhibit a significantly lower forward voltage drop, typically around 0.2-0.4 volts, compared to 0.7 volts for silicon diodes. This translates to lower power loss during forward conduction.

- **Fast switching speed:** Due to the absence of a p-n junction and the minimal storage time for charge carriers, Schottky diodes offer very fast switching speeds, making them ideal for high-frequency applications[12,13,14].

- **Leakage current:** While generally lower than standard diodes, Schottky diodes do exhibit a higher reverse leakage current. This means a small current still flows even when the diode is reverse biased.

2. Applications

- **Power supply rectification:** Due to their low forward voltage drop and high efficiency, Schottky diodes are often used as rectifiers in power supplies, especially for low-voltage applications like battery chargers and voltage regulators[15,16,17].

- **High-frequency applications:** Their fast-switching speeds make Schottky diodes suitable for various high-frequency circuits, including mixers, detectors, and RF switches.

- **Logic circuits:** In low-power logic families like Schottky TTL (Transistor-Transistor Logic), these diodes are used to improve speed and reduce power consumption.

- **Voltage clamping:** They can be used for clamping voltage spikes or transients in circuits to protect sensitive components.

3. Comparison to standard p-n junction diodes

While Schottky diodes offer advantages like lower forward voltage drop and faster switching, they also have drawbacks:

- **Higher reverse leakage current:** As mentioned earlier, Schottky diodes have a higher reverse leakage current compared to standard p-n junction diodes. This can be a concern in applications where leakage current needs to be minimal.

- **Lower breakdown voltage:** Generally, Schottky diodes have a lower breakdown voltage compared to p-n junction diodes. This means they can withstand lower reverse voltages before breaking down.

- **Temperature sensitivity:** The forward voltage drop of a Schottky diode increases with temperature more significantly than that of a p-n junction diode. This needs to be considered in applications where temperature variations are high[5,8].

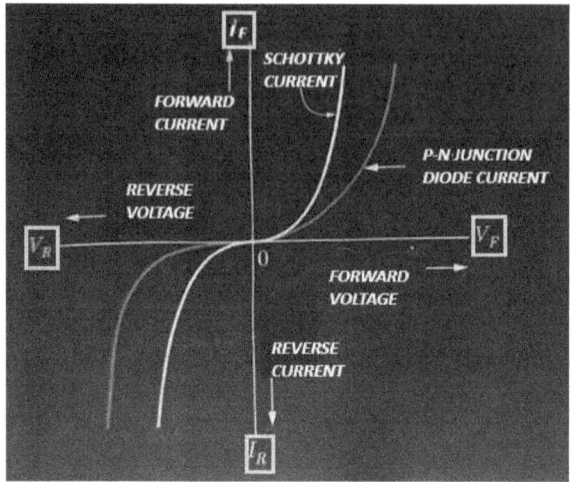

Figure 12 I-V Characteristics of Schottky diodes.

Chapter-3

Bipolar Junction Transistor (BJT)

3.1 What is a Transistor?

- **A** transistor is a fundamental semiconductor device used for two primary purposes:
    - **Amplification:** Transistors can amplify weak electronic signals to create a stronger output signal.
    - **Switching:** Transistors can act as tiny switches, rapidly turning current flow on and off.
- Transistors form the building blocks of integrated circuits (ICs), which are the foundation of modern electronic devices like computers, smartphones, and countless others[14].

3.2 Types of Transistors

The two main categories of transistors are:

3.2.1 Bipolar Junction Transistors (BJTs)

- Constructed with three-layers of doped semiconductor material (N-P-N or P-N-P configurations).
- Current flow is controlled by two types of charge carriers: electrons and holes.
- Operation: A small current (or voltage) applied to the "base" terminal controls a much larger current flow between the "collector" and "emitter" terminals[17].

3.2.2 Field-Effect Transistors (FETs)

- Constructed using a single type of semiconductor material, either p-type or n-type.
- Current flow is controlled by only one type of charge carrier: electrons (in an N-channel FET) or holes (in a P-channel FET).
- Operation: An electric field generated by a voltage on the "gate" terminal controls the conductivity of a channel between the "source" and "drain" terminals[15,19].

1. Subtypes of FETs:

- **JFET (Junction Field-Effect Transistor):** Contains a p-n junction that forms the gate.
- **MOSFET (Metal-Oxide-Semiconductor Field-Effect Transistor):** The gate is formed by a metal oxide layer placed on top of the semiconductor channel. MOSFETs are the most commonly used transistors in modern electronics due to their small size, high efficiency, and versatility[20,25].

Feature	Bipolar Junction Transistor (BJT)	Field-Effect Transistor (FET)
Construction	Three layers of doped semiconductor	Single channel of doped semiconductor
Charge carriers	Electrons and holes	Electrons (N-channel) or holes (P-channel)
Current control	Base current controls collector-emitter current	Gate voltage controls source-drain current
Input Impedance	Relatively low	Very high

Transconductance	Higher than FETs (more current output for a given input change)	Generally lower than BJTs
Power Consumption	Generally higher than FETs	Typically lower than BJTs
Applications	Power amplification circuits, some amplifiers, specific switching applications	Digital ICs, low-power circuits, switches, RF circuits

Table 3 BJT VS FET

3.3 Basic Structure of a Bipolar Junction Transistor (BJT)

A BJT, or Bipolar Junction Transistor, is a three-terminal semiconductor device that allows for amplification and switching of electrical signals. Here's a breakdown of its basic structure[1,2,8]:

Components:

- **Semiconductor Material:** BJTs are typically made from silicon, doped with impurities to create specific regions:

    o **Base (B):** The middle region, typically lightly doped to allow for easy movement of charge carriers.

    o **Emitter (E):** The source of charge carriers, heavily doped to easily inject them into the base[5,11].

    o **Collector (C):** The region that collects the majority of the injected charge carriers, moderately doped for efficient collection.

- **p-n Junctions:** Two p-n junctions exist within the device:

- **Base-Emitter Junction (BE Junction):** Separates the base and emitter regions.
- **Base-Collector Junction (BC Junction):** Separates the base and collector regions.

Doping Types:

- **NPN Transistor:** The most common type, with an n-type emitter, a p-type base, and an n-type collector.
- **PNP Transistor:** Less common, with a p-type emitter, an n-type base, and a p-type collector.

Terminals:

- **Emitter (E):** Connected to the most heavily doped region, where charge carriers are injected.
- **Base (B):** Controls the flow of charge carriers between emitter and collector[2].
- **Collector (C):** Collects the majority of the injected charge carriers and outputs the amplified current.

3.4 Typical Doping Concentrations for BJT

In a Bipolar Junction Transistor (BJT), the doping concentrations in its different regions (emitter, base, and collector) play a crucial role in determining its electrical characteristics and performance. Here's an explanation of the typical doping concentrations for a BJT graph[1,2,3,4,5]:

1. **Emitter Region:**
   - The emitter region is heavily doped to ensure efficient injection of majority carriers (electrons in an NPN transistor or holes in a PNP transistor) into the base region.

- Typically, the doping concentration in the emitter region is several orders of magnitude higher than that of the base and collector regions.

- The doping concentration in the emitter region is represented by a steep incline or peak in the graph, indicating a high density of dopant atoms.

2. **Base Region:**

    - The base region is lightly doped to allow for the diffusion of minority carriers (holes in an NPN transistor or electrons in a PNP transistor) injected from the emitter region to reach the collector region.

    - The doping concentration in the base region is significantly lower than that of the emitter region but higher than that of the collector region.

    - The base doping profile typically shows a gradual decrease from the heavily doped emitter junction to the lightly doped collector junction. This gradient facilitates the transit of minority carriers across the base region[9,12].

3. **Collector Region:**

    - The collector region is moderately doped to collect the majority carriers (electrons in an NPN transistor or holes in a PNP transistor) from the base region and allow them to flow to the external circuit.

    - The doping concentration in the collector region is lower than that of the emitter region but higher than that of the base region.

    - The collector doping profile exhibits a relatively uniform distribution with a concentration lower than that of the emitter but higher than that of the base[22,25,26].

Overall, the doping concentrations graph for a BJT typically shows a sharp peak for the emitter, a gradual decrease for the base, and a moderate, uniform concentration for the collector. This doping profile is designed to optimize the transistor's operation for amplification and switching applications while ensuring efficient carrier transport across its different regions[21].

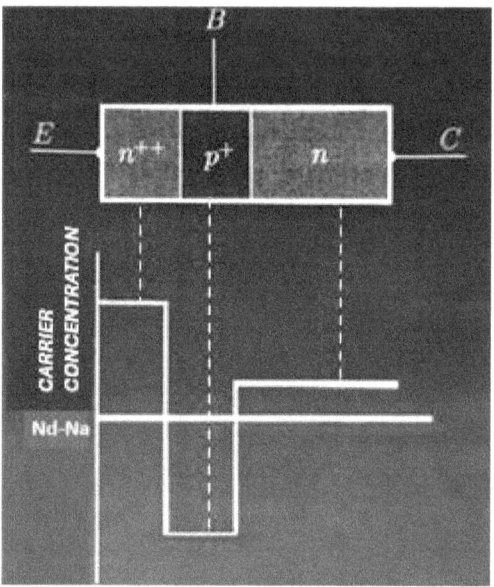

Figure 13 Doping concentrations graph for a BJT.

3.5 Transistor Biasing

The three main biasing configurations for bipolar junction transistors (BJTs) are:

1. **Common Emitter Configuration (Forward Active Bias):**

    - In this configuration, the emitter-base junction is forward biased, while the collector-base junction is reverse biased.

    - Forward biasing of the emitter-base junction allows a constant forward current to flow from the emitter to the base, controlling the transistor's operation[25].

- Reverse biasing of the collector-base junction creates a depletion region, preventing significant current flow from the collector to the base.

- This configuration provides high current gain (β) and moderate input and output impedance, making it suitable for amplification purposes[24].

2. **Common Base Configuration (Reverse Active Bias):**

    - In the common base configuration, the base is grounded, and the input signal is applied to the emitter terminal[22].

    - The emitter-base junction is forward biased, allowing input signal current to flow into the base.

    - The collector-base junction is reverse biased, allowing the transistor to operate in the reverse active region[24].

    - This configuration offers low input impedance, high voltage gain, and wide bandwidth, making it suitable for impedance matching and high-frequency applications[5,11,14].

3. **Common Collector Configuration (Emitter Follower or Voltage Buffer):**

    - In the common collector configuration, also known as the emitter follower or voltage buffer configuration, the collector is connected to the input voltage source, and the output is taken from the emitter.

    - The emitter-base junction is forward biased, while the collector-base junction is reverse biased.

    - This configuration offers unity voltage gain, high input impedance, and low output impedance, making it ideal for impedance matching and buffering applications.

- It provides a high degree of signal isolation between the input and output, making it useful for driving loads with minimal signal distortion.

These biasing configurations play a crucial role in determining the transistor's operating characteristics and performance in different circuit applications. The choice of biasing configuration depends on the specific requirements of the circuit, such as gain, impedance matching, bandwidth, and signal isolation[17].

3.6 Transistor Operation

Feature	Active Region	Saturation Region	Cutoff Region	Inverse Active Region
Purpose	Amplification of signals	Electronic switching	No current flow, transistor acts as off	Rarely used, similar to active, but with roles of collector and emitter reversed
Biasing	Base-emitter junction forward-biased Base-collector junction reverse-biased	Both junctions forward-biased	Both junctions reverse-biased	Base-emitter junction reverse-biased. Base-collector junction forward-biased

Feature	Active Region	Saturation Region	Cutoff Region	Inverse Active Region
Current Behavior	Collector current (Ic) is proportional to base current (Ib) **Key Equation:** Ic = β * Ib (β is the current gain)	Collector current (Ic) maximum, no longer proportional to Ib. Base-collector junction acts as short circuit.	Collector and emitter current are essentially zero.	Similar to active region, but current flows in reverse direction (emitter to collector.) Less efficient.
Voltage Behavior	Vce (Collector-Emitter voltage) is greater than Vbe (Base-Emitter voltage) but not high enough to maximize Ic.	Vce is very small, approximately 0.2V. Transistor acts "on" like a closed switch.	Vce can be large, as no current flows	Vce is similar to that of the active region

Table 4 Transistor Operations regions

E-B Junction	C-B Junction	Region of Operation
FB	RB	Active
FB	FB	Saturation
RB	RB	Cut-off
RB	FB	Inverse

Table 5 Bias of transistor operation regions

3.7 Current Components in a Transistor

As a result of Biasing in the active region, current flows due to drift and diffusion in various parts of the transistor as highlighted below[1,8].

$$I_E : Emitter\ current$$

Where, I_{PE}, I_{NE} : *Emitter current (P: Hole component), (N: Electron Component)*

$$I_p = \frac{AQDP}{Lp}(e^{V/VT}-1)$$

$$Thus\ I_{pE} = \frac{AQDP}{w}(e^{VEB/VT}-1)$$

$$I_{NE} = \frac{AQDN}{w}(e^{VEB/VT}-1)$$

The Hole component is responsible for the transistor action and $I_{PN} \gg I_{PE}$.

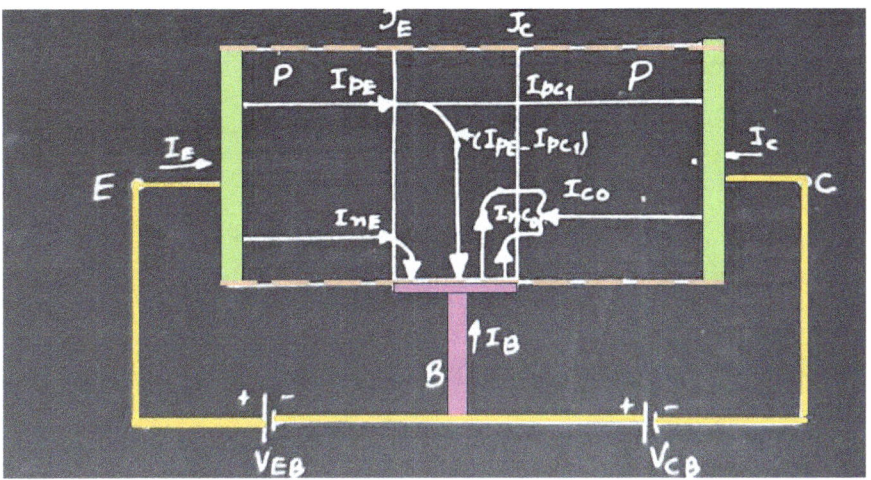

Figure 14 Current components of a transistor.

3.8 Transistor characterization

Current Relationships:

Emitter Current (IE): This is the total current entering the transistor through the emitter terminal. It's the sum of base current (IB) and collector current (IC)[15].

$$Equation:\ IE = IB + IC$$

Collector Current (IC): This is the main current output of the transistor, flowing from the collector to the emitter. It's related to the base current by a factor called beta (β).

$$Equation:\ IC = \beta * IB$$

(This is the most common equation, but alpha (α) can also be used, see below)

Base Current (IB): This is a small control current entering the transistor through the base terminal. It controls the much larger collector current.

While not as commonly used for BJT[18] characterization, here's the equation for completeness:

$$IB = (IE - IC)$$

Current Gain Factors:

Beta (β): This is the most common current gain factor in BJTs, representing the ratio of collector current to base current in the active region.

$$\text{Equation: } \beta = IC / IB$$

Alpha (α): This is another current gain factor, representing the ratio of collector current to emitter current. It's slightly less common than beta but still relevant.

$$\text{Equation: } \alpha = IC / IE$$

(Relationship between alpha and beta)

$$\beta = \alpha / (1 - \alpha)$$

Note: Beta is typically a much larger value than alpha (usually in the range of 50-200).

Additional Equations (Depending on Model):

1. Shockley Diode Equation (for forward-biased base-emitter junction): This equation describes the exponential relationship between the base-emitter voltage (VBE) and the base current (IB) but is a more complex model than the basic IB = (IE - IC) used for characterization[22].

2. Early Effect: This effect describes how the collector current can also be influenced by the collector-emitter voltage (VCE) at higher voltages. It's usually a secondary effect for basic BJT characterization.

3. **Temperature Dependence:** The Shockley Diode Equation describes the exponential relationship between base-emitter voltage (VBE) and base current (IB). To include temperature dependence, we introduce thermal voltage (VT = kT/q)

4. **Modified Shockley Diode Equation:**

$$I_E = I_S \cdot \left(\exp\left(\frac{V_{BE}}{V_T}\right) - 1 \right)$$

where Is: Saturation current (dependent on temperature)

k: Boltzmann's constant (1.381 x 10^-23 J/K)

T: Absolute temperature (in Kelvin)

q: Electron charge (1.602 x 10^-19 C)

5. **Ebers-Moll Model:**

This advanced model considers leakage currents and other effects. It uses a system of equations with matrix notation to relate all terminal currents (IE, IB, IC) to terminal voltages (VBE, VCE). Here's a simplified representation[14,15,21,22]

Collector Current (IC):

$$I_C = (\alpha \cdot I_E) + \left(J_{SV} \cdot \left(1 - \exp\left(\frac{V_{BE}}{V_T}\right) \right) \right) + \left(J_{SC} \cdot \left(1 - \exp\left(\frac{V_{CE}}{V_T}\right) \right) \right)$$

Where α: Current gain factor (similar to but not exactly equal to beta)

J_{SV}, J_{SC}: Reverse saturation current densities for base-emitter and collector-emitter junctions

Emitter Current (IE) and Base Current (IB) can be expressed similarly with additional terms for leakage currents.

Specific Device Characterization:

For a specific BJT device, you might introduce parameters like Early voltage (VA) which relates to the collector current dependence on collector-emitter voltage.[15]

6. **Collector Current with Early Effect:**

$$I_C = I_S \cdot \left(\exp\left(\frac{V_{BE}}{V_T}\right) - 1\right) + \left(\frac{V_{CE}}{V_A}\right) \cdot I_E$$

where VA: Early voltage (device-specific parameter)

3.8.1. Circuit-Level Analysis with Detailed BJT Models

Detailed BJT models can be used for small-signal analysis of BJT amplifiers. Techniques like S-parameters involve complex number representations of voltage and current phasors. The specific equations depend on the chosen analysis method[17,18,20].

While basic BJT models (e.g., hybrid-pi model) are sufficient for initial analysis, detailed BJT models offer a more accurate picture, especially for high-frequency circuits. Here's how detailed models are used in small-signal analysis with examples:

Case Study 1: Common-Emitter Amplifier with Ebers-Moll Model

- **Scenario:** We want to analyze the small-signal gain and frequency response of a common-emitter amplifier using the Ebers-Moll model[21].

- **Detailed Model:** The Ebers-Moll model considers base-emitter and collector-emitter junction leakage currents, providing a more comprehensive description than the simpler hybrid-pi model[18].

- **Analysis Method:** S-parameter analysis is a common technique for small-signal analysis of transistor circuits. It involves converting voltages and currents into their phasor counterparts using complex numbers[19].

- **Equations:** The Ebers-Moll model equations are transformed into the S-parameter domain using matrix manipulations. Software tools can be used to perform these calculations.

- **Benefits:** The S-parameters obtained from the detailed model capture the amplifier's gain, input and output impedance, and phase response across a range of frequencies. This provides a more accurate prediction of the amplifier's behavior compared to a simpler model.

Case Study 2: High-Frequency BJT Amplifier with Gummel-Poon Model

- **Scenario:** We're designing a high-frequency BJT amplifier and need to consider the effects of transit time (the time it takes carriers to travel across the base region)[15].

- **Detailed Model:** The Gummel-Poon model incorporates transit time and other high-frequency effects not captured by the Ebers-Moll model[22,24].

- **Analysis Method:** Similar to the previous case study, S-parameter analysis can be used with the Gummel-Poon model's equations transformed into the S-parameter domain. Alternatively, high-frequency analysis techniques like h-parameter equivalent circuits with frequency-dependent elements can be employed.

- **Benefits:** The Gummel-Poon model provides a more accurate prediction of the amplifier's gain roll-off and frequency limitations compared to simpler models. This is crucial for designing high-frequency circuits.

Additional Considerations:

- **Computational Complexity:** Detailed models like Ebers-Moll and Gummel-Poon involve more complex equations, leading to increased computational demands compared to simpler models. SPICE simulation tools can be used to efficiently analyze circuits with these models[25].

- **Model Parameter Extraction:** Extracting accurate model parameters for detailed BJT models often requires specialized measurement techniques.

3.9 Transistors at low frequency: Characteristics and Analysis

At low frequencies, transistors behave in a more predictable manner compared to high frequencies. This allows for simpler analysis using basic transistor models. Here are some key points to consider[15]:

1. Biasing and Operating Regions:

- **DC biasing:** Transistors require proper DC biasing to operate in the desired region (active, saturation, cutoff). Biasing sets the base-emitter (VBE) and collector-emitter (VCE) voltages[18].

- **Active region:** This is the primary region for amplification. In this region, the base-emitter junction is forward-biased and the collector-base junction is reverse-biased[21].

2. Small-Signal Analysis:

- **Small-signal model:** Low-frequency analysis often utilizes small-signal models like the hybrid-pi model. This model represents the transistor using capacitors, resistors, and a current source to capture its small-signal AC behavior around a specific DC operating point[24].

- **h-parameters:** These are small-signal parameters of the hybrid-pi model (e.g., hi, hfe, hro) and relate small-signal changes in voltage and current at the transistor terminals.

3. Mathematical Analysis with h-parameters:

- **Voltage Gain (Av):**

$$A_v = -\frac{h_{fe} \cdot R_L}{R_e + \left(\frac{h_i \cdot h_{ie}}{h_i + h_{ie}}\right)}$$

where,

  o hfe: Small-signal current gain

  o hi: Input impedance of the transistor model.

- o hie: Emitter resistance of the transistor model (often neglected at low frequencies)
- o ||: Represents parallel resistance.
- **Input Impedance (Zin):** *Zin ≈ Re + (hi || hie)*
- **Output Impedance (Zout):** *Zout ≈ hro || RL* (hro: Output impedance of the transistor model)

These equations allow you to calculate the voltage gain, input impedance, and output impedance of the amplifier at low frequencies. They provide valuable insights into the amplifier's performance, such as its amplification capability and signal source loading effect[14,17].

4. Limitations at Low Frequencies:

- **Capacitive effects:** While neglected in basic low-frequency analysis, capacitors within the transistor model become more significant at higher frequencies and need to be considered for accurate analysis.

3.9.1 Hybrid Model for CB Configurations

The hybrid model for the CB configuration is typically used to analyze the behavior of bipolar junction transistors (BJTs) in this configuration. It combines the small-signal equivalent circuit model with the large-signal (nonlinear) characteristics of the transistor[7,8,9].

The small-signal hybrid-π model for the CB configuration consists of parameters such as the hybrid π-model conductance gm, $g\pi$, go, and gr, where:

- *gm* represents the transconductance,
- *gπ* represents the base-spreading conductance,
- *go* represents the output conductance, and
- *gr* represents the reverse transfer conductance.

The equations defining the small-signal hybrid-π model for the CB configuration can be expressed as:

$$i_c = g_m \cdot v_{be} - g_\pi \cdot v_\pi + g_o \cdot v_{cb} - g_r \cdot v_{rc}$$

$$i_b = (g_m + g_\pi) \cdot v_{be} - g_\pi \cdot v_\pi + g_r \cdot v_{rc}$$

$$i_e = i_c + i_b$$

Here, *ic* is the collector current, *ib* is the base current, *ie* is the emitter current, *vbe* is the voltage across the base-emitter junction, vπ is the voltage across the base-spreading resistance, vcb is the voltage across the collector-base junction, and *vrc* is the voltage across the reverse-biased collector-base junction[11,12].

3.9.2 Hybrid Model for CE Configuration

The common emitter (CE) configuration is one of the most commonly used configurations in bipolar junction transistor (BJT) circuits. The hybrid model for the CE configuration combines small-signal equivalent circuit models with the large-signal characteristics of the transistor. This model allows for the analysis of the CE amplifier's behavior under small-signal conditions, making it valuable in designing and analyzing amplifier circuits.

Small-Signal Hybrid-π Model:

The small-signal hybrid-π model for the CE configuration consists of parameters such as the hybrid π-model conductance *gm*, *gπ*, *go*, and *gr*, where:

- *gm* represents the transconductance,
- *gπ* represents the base-spreading conductance,
- *go* represents the output conductance, and
- *gr* represents the reverse transfer conductance.

The equations defining the small-signal hybrid-π model for the CE configuration are as follows:

For *ic*:

$$i_c = g_m \cdot v_{be} - g_\pi \cdot v_\pi + g_o \cdot v_{cb} - g_r \cdot v_{rc}$$

For *ib*:

$$i_b = (g_m + g_\pi) \cdot v_{be} - g_\pi \cdot v_\pi + g_r \cdot v_{rc}$$

For *ie*:

$$i_e = i_c + i_b$$

3.9.3 Hybrid Model for CC Configuration

The common collector (CC) configuration, also known as the emitter follower, is a widely used configuration in bipolar junction transistor (BJT) circuits. The hybrid model for the CC configuration combines small-signal equivalent circuit models with the large-signal characteristics of the transistor. This model allows for the analysis of the CC amplifier's behavior under small-signal conditions, making it valuable in designing and analyzing amplifier circuits[11].

Small-Signal Hybrid-π Model:

The small-signal hybrid-π model for the CC configuration consists of parameters such as the hybrid π-model conductances *gm*, *gπ*, *go*, and *gr*, where:

- *gm* represents the transconductance,
- *gπ* represents the base-spreading conductance,
- *go* represents the output conductance, and
- *gr* represents the reverse transfer conductance.

$$i_e = g_m \cdot v_{be} - g_\pi \cdot v_\pi + g_o \cdot v_{cb} - g_r \cdot v_{rc}$$
$$i_b = (g_m + g_\pi) \cdot v_{be} - g_\pi \cdot v_\pi + g_r \cdot v_{rc}$$

$$i_c = i_e - i_b$$

3.9.4 Questions based on Design and Simulation

Q: Question, A: Answer

3.9.4.1 Circuit Design and Analysis

Q: Design a low-frequency BJT amplifier with specific gain and impedance requirements, considering distortion and stability.

A:

- **Design Approach:**
  - Choose BJT type (e.g., bipolar junction transistor) based on gain and frequency requirements.
  - Select appropriate biasing resistors for desired operating point and gain.
  - Design the collector load resistor to achieve the target voltage gain.
  - Consider using bypass capacitors to minimize bias voltage variations across internal transistor resistances.
  - Employ feedback techniques (e.g., negative feedback) for improved linearity and reduced distortion.

- **Component Selection:**
  - Select BJTs with appropriate hfe for desired gain.
  - Choose resistors with tight tolerances for stable biasing.
  - Select capacitors with low leakage current and values suitable for the desired low-frequency response.

- **Analysis Techniques:**

- **h-parameter analysis:** Calculate gain, input/output impedance using h-parameters and circuit equations.
- **SPICE simulation:** Simulate the designed circuit to verify gain, frequency response, distortion levels, and stability across the operating range.

Q: Modify a BJT amplifier design for wide frequency response and good low-frequency gain.

A:

- **Techniques:**

- Use Miller feedback capacitors to compensate for the Miller effect and extend the high-frequency response.
- Bootstrap techniques can further improve high-frequency gain.
- Careful selection of biasing resistors ensures good low-frequency gain is maintained.

- **Trade-offs:**

- Increased capacitance might introduce lower roll-off at low frequencies.
- Feedback can reduce gain slightly.

- **Stability Considerations:**

- Proper phase margin in the frequency response ensures stability.
- SPICE simulations can be used to verify stability across the desired frequency range.

3.9.4.2. Advanced Modeling and Analysis:

Q: Explain limitations of the hybrid-pi model and how the Gummel-Poon model addresses them.

A:

- **Hybrid-pi Limitations:**

  - Neglects base width modulation (Early effect) which affects collector current at higher voltages.

  - Ignores high-frequency effects like transit time and collector junction capacitance.

- **Gummel-Poon Model:**

  - Includes Early effect for more accurate collector current modeling.

  - Models base width modulation and collector junction capacitance, improving high-frequency analysis.

Q: Model and analyze thermal behavior of a BJT amplifier at low frequencies.

A:

- **Modeling:**

  - Develop a thermal model that considers power dissipation in the transistor and heat transfer to the surrounding environment.

  - Include temperature dependence of transistor parameters (e.g., hfe, Early voltage) in the model.

- **Analysis:**

  - Use SPICE simulation with thermal models to analyze how self-heating affects amplifier performance (gain, stability) over time.

  - Design considerations include proper heatsinking and biasing for thermal stability.

3.9.4.3 Real-World Applications and Troubleshooting

Q: Troubleshoot a BJT amplifier with unexpected gain fluctuations.

A:

- **Troubleshooting Approach:**

  - Verify power supply voltages for stability.

  - Measure biasing currents and voltages to identify potential biasing resistor drift.

  - Check for component failures (e.g., open resistors, leaky capacitors) using a multimeter.

  - Analyze temperature using thermal imaging to identify potential overheating issues.

Q: Design a low-frequency BJT amplifier robust against power supply and temperature variations.

A:

- **Design Techniques:**

  - Use constant current sources for biasing instead of resistors to reduce dependence on power supply variations.

    - Select components with good temperature coefficients to minimize parameter variations with temperature.

    - Thermal compensation techniques (e.g., using a temperature-dependent resistor) can further improve stability.

3.9.4.4 Leadership and Problem-Solving:

Q: Lead a team developing a low-noise BJT amplifier for a medical device.

A:

- **Project Management:**

- Clearly define project goals, performance requirements, and deadlines.
- Delegate tasks based on team member expertise.
- Foster open communication and regular progress reviews.
  - **Problem-Solving Strategies:**
- Encourage brainstorming sessions to identify potential solutions.
- Facilitate data analysis and interpretation from simulations and experiments.
- Guide the team towards selecting the optimal solution based on performance, manufacturability, and regulatory compliance.

3.10 Transistors at high frequency

When operating transistors at high frequencies, certain effects come into play that significantly impact their performance. These effects include parasitic capacitances, transit time effects, and the limitations imposed by the device's intrinsic characteristics. Understanding and mitigating these effects are crucial for designing high-frequency circuits accurately[22,24,26].

1. Parasitic Capacitances:

- At high frequencies, the parasitic capacitances associated with transistors become significant. These include the junction capacitances (C_{jc}, C_{je}) and the capacitances between the terminals (e.g., C_{bc}, C_{be}).
- These capacitances can affect the transistor's gain, impedance, and bandwidth, leading to deviations from the desired behavior[1,2,8,10].

2. Transit Time Effects:

- In high-frequency operation, the time taken for carriers to cross the transistor's base region becomes non-negligible, known as transit time.

- Transit time leads to phase delay and distortion in the transistor's response, impacting its frequency response and stability[8,9,12,15].

3. Intrinsic Device Limitations:

- Intrinsic device limitations, such as the Early effect in BJTs and the output conductance in MOSFETs, impose restrictions on high-frequency performance.

- The Early effect causes a variation in the transistor's collector current with collector voltage, affecting its gain and bandwidth.

- Output conductance introduces a conductive path between drain and source, reducing gain and bandwidth in MOSFETs.

Mathematical Equations:

1. High-Frequency Small-Signal Model:

For BJTs, the high-frequency small-signal model incorporates the parasitic capacitances and transit time effects. The hybrid-$\pi\pi$ model is extended to include these effects[13,15]. Mathematically, the high-frequency hybrid-$\pi\pi$ model for a BJT can be represented as:

$$i_c = g_m v_{be} - g_\pi v_\pi + g_o v_{cb} - g_r v_{rc}$$

$$i_b = (g_m + g_\pi) v_{be} - g_\pi v_\pi + g_r v_{rc}$$

$$i_e = i_c + i_b$$

- ic is the collector current,
- ib is the base current,
- ie is the emitter current,
- vbe is the voltage across the base-emitter junction,
- $v\pi$ is the voltage across the base-spreading resistance,

- *vcb* is the voltage across the collector-base junction,
- *vrc* is the voltage across the reverse-biased collector-base junction,
- *gm* is the transconductance,
- *gπ* is the base-spreading conductance,
- *go* is the output conductance, and
- *gr* is the reverse transfer conductance.

2. MOSFET Model:

For MOSFETs, the high-frequency small-signal model involves the incorporation of parasitic capacitances and output conductance. The small-signal model includes these effects along with the intrinsic capacitances (gate-source and gate-drain capacitances) and channel conductance. Mathematically representing this model involves similar equations as for BJTs but adapted for MOSFET characteristics[14,18].

In summary, at high frequencies, transistors exhibit deviations from ideal behavior due to parasitic capacitances, transit time effects, and intrinsic device limitations. Understanding these effects and incorporating them into mathematical models is essential for designing high-frequency circuits accurately.

The Metal-Oxide-Semiconductor Field-Effect Transistor (MOSFET) is a fundamental building block in modern integrated circuits and electronic devices. Understanding its behavior involves a small-signal model that encapsulates various parameters and effects. Here's an explanation of the MOSFET model along with equations[19,20]:

1. Basic MOSFET Structure:

A MOSFET consists of a gate, source, drain, and substrate (bulk). The channel between the source and drain is controlled by the gate voltage[15,18].

Based on the structure, MOSFETs are classified into two main types: Enhancement-mode (E-MOSFET) and Depletion-mode (D-MOSFET).

2. Small-Signal Model:

The small-signal model of a MOSFET is commonly expressed using the transconductance ($gmgm$), output conductance (gd), and capacitances (Cgs, Cgd).

3. Equations:

3.1 Transconductance (gm):

The transconductance (gm) represents the change in drain current (id) with respect to the change in gate-source voltage (vgs) at a fixed drain-source voltage (vds). It is given by:

$$g_m = \frac{\partial i_d}{\partial v_{gs}}$$

3.2 Output Conductance (gd):

The output conductance (gd) represents the change in drain current (id) with respect to the change in drain-source voltage (vds) at a fixed gate-source voltage (vgs). It is given by:

$$g_d = \frac{\partial i_d}{\partial v_{ds}}$$

3.3 Capacitances (Cgs and Cgd):

The capacitance between the gate and source (Cgs) and between the gate and drain (Cgd) are important parameters affecting the MOSFET's high-frequency behavior. They are defined by[16,19]:

$$C_{gs} = \frac{\partial q_{gs}}{\partial v_{gs}}$$

$$C_{gd} = \frac{\partial q_{gd}}{\partial v_{gd}}$$

3.10.1 Validity and Parameter Variation

At high frequencies, the validity of the CE model may be compromised by factors such as parasitic elements, stray capacitances, and non-linear effects. The equations describing parameter variations due to changes in bias conditions, temperature, and fabrication process become more complex and may involve differential equations or nonlinear functions[12,14,8].

Parameter Variations Equations:

Small-Signal Transconductance (*gm*) Variation: At high frequencies, variations in bias conditions and temperature can affect the small-signal transconductance (*gmgm*) of the transistor. This variation can be modeled using a nonlinear function:

$$[g_m = g_{m0} \cdot f(V_{BE}, T)]$$

where g_{m0} is the small-signal transconductance at a reference bias and temperature, and $f(V_{BE}, T)$ is a function that describes how g_m varies with changes in the base-emitter voltage (V_{BE}) and temperature (T).

Output Conductance (*gogo*) Variation: Similarly, variations in bias conditions and temperature can impact the output conductance (*gogo*) of the transistor. This variation can also be modeled using a nonlinear function:

$$[g_o = g_{o0} \cdot g(V_{CE}, I_C, T)]$$

where g_{o0} is the output conductance at a reference bias and temperature, and $g(V_{CE}, I_C, T)$ is a function that describes how g_o varies with changes in the collector-emitter voltage (V_{CE}), collector current (I_C), and temperature (T).

Base Transit Time Variation: Changes in bias conditions and temperature can also affect the base transit time of the transistor, which influences its high-frequency behavior. This variation can be expressed as:

$$[\tau_b = \tau_{b0} \cdot h(V_{BE}, T)]$$

where τb0 is the base transit time at a reference bias and temperature, and h(VBE,T) is a function that describes how τb varies with changes in the base-emitter voltage (VBE) and temperature (T).

Nonlinear Function Equations[22]:

Here are examples of nonlinear functions that describe the variations:

1. **Temperature Dependence Function:**

$$[f(T) = e^{(T-T_0)/T_0}]$$

2. **Voltage and Current Dependence Function:**

$$[g(V,I,T) = \frac{V^2}{1 + e^{-I/T}}]$$

3. **Combined Voltage and Temperature Dependence Function:**

$$[h(V,T) = \sqrt{V} + e^{(T-T_0)/T_0}]$$

These equations and functions capture the complex variations in transistor parameters due to changes in bias conditions, temperature, and fabrication process, providing insights into the challenges associated with high-frequency transistor modeling.

3.10.2 Current gain High frequency BJT vs Low frequency BJT

The current gain (β) of a bipolar junction transistor (BJT) is a crucial parameter that quantifies the ratio of collector current (IC) to base current (IB). However, the behavior of β can vary significantly between high-frequency and low-frequency regimes due to the influence of parasitic capacitances, transit time effects, and other factors. Let's compare the mathematical equations for β in high-frequency (HF) and low-frequency (LF) BJTs[15]:

Low-Frequency BJT:

In the low-frequency regime, the base transit time (τb) and parasitic capacitances ($C\pi$ and $C\mu$) have negligible effects on the transistor behavior. Therefore, the current gain β can be accurately approximated by the simplified expression:

$$[\beta_{LF} = \frac{I_C}{I_B}]$$

Where:

- IC is the collector current.
- IB is the base current.

This equation assumes ideal transistor behavior and neglects the influence of parasitic elements, making it suitable for low-frequency analysis[18].

High-Frequency BJT:

In the high-frequency regime, the presence of parasitic capacitances ($C\pi$ and $C\mu$) and the base transit time (τb) significantly affect the transistor's dynamic behavior. As a result, the current[21] gain β in high-frequency BJTs is more accurately described by the hybrid-$\pi\pi$ model and can be expressed as:

$$[\beta_{HF} = \frac{\frac{g_m}{s}}{C_\pi + C_\mu}]$$

Where:

- gm is the transconductance of the transistor.
- s is the Laplace variable representing the frequency domain.
- $C\pi$ is the junction capacitance between the base and the collector.

- $C\mu$ is the junction capacitance between the base and the emitter.

In high-frequency analysis, β becomes a frequency-dependent parameter due to the presence of parasitic capacitances. The dynamic behavior of the transistor is captured by the Laplace variable s, which accounts for the frequency-dependent response[22].

Comparison:

- In low-frequency BJTs, β is constant and independent of frequency, making it easier to analyze and design circuits.
- In high-frequency BJTs, β becomes frequency-dependent due to the influence of parasitic elements, requiring more complex models for accurate analysis and design[25].

3.10.3 Current gain with resistive load

The current gain with resistive loads in a bipolar junction transistor (BJT) circuit refers to the amplification of input current to output current when the transistor is connected to resistive loads. This gain is commonly denoted as β_{RL}.

Explanation:

When a BJT is connected to a resistive load (RL), the collector current (I_C) flows through the load resistor, causing a voltage drop across it. This voltage drop results in an output voltage signal. The base current (IB) controls the transistor's conductivity, which in turn affects IC. Therefore, the current gain with resistive loads can be expressed as the ratio of the change in collector current to the change in base current:

$$[\beta_{RL} = \frac{\Delta I_C}{\Delta I_B}]$$

This equation quantifies the amplification effect of the transistor with resistive loads. A higher value of β_{RL} indicates greater amplification capability.

When designing BJT circuits with resistive loads, understanding the current gain (β_{RL}) is crucial for determining the amplification and signal processing capabilities of the circuit.

3.10.4 Frequency response of single stage CE Amplifier

The frequency response of a single-stage common emitter (CE) amplifier refers to how its gain varies with the frequency of the input signal. This response is influenced by the transistor's internal capacitances, resistances, and other circuit components[18].

Explanation:

In a CE amplifier, the gain is typically calculated as the ratio of the output voltage to the input voltage. At low frequencies, the gain is primarily determined by the transistor's DC biasing conditions and the external circuit components. However, as the frequency increases, the behavior of the transistor's internal capacitances becomes significant, affecting the amplifier's performance[15].

The key factors affecting the frequency response of a single-stage CE amplifier are:

1. **Internal Capacitances:** The transistor's internal capacitances, such as the junction capacitances ($C\pi$ and $C\mu$), influence the amplifier's gain at higher frequencies. These capacitances introduce impedance that interacts with the load and input impedance of the amplifier, affecting its gain[22].

2. **Miller Effect:** The Miller effect occurs when the input and output capacitances of the transistor form a voltage divider, reducing the effective gain of the amplifier at high frequencies. This effect is particularly significant in CE amplifiers due to the configuration of the transistor[21].

Mathematical Equations:

The frequency response of a single-stage CE amplifier can be represented mathematically using various circuit analysis techniques, such as small-signal analysis and AC equivalent circuits. One common approach is to derive the amplifier's transfer function, which relates the output voltage ($Vout$) to the input voltage (Vin) as a function of frequency (f)[18].

3.10.5 Definition of GAIN- BANDWIDTH PRODUCT

The gain-bandwidth product (GBP) is a key parameter in amplifier design that represents the product of the amplifier's gain and its bandwidth. It indicates the frequency range over which a constant gain can be maintained. GBP is particularly important in operational amplifiers and other high-frequency circuits where maintaining a consistent gain across a wide range of frequencies is essential[22].

Explanation:

The GBP is defined as the product of the open-loop voltage gain ($AvAv$) and the bandwidth ($f3dBf3dB$) of the amplifier. Mathematically, it can be expressed as:

$$\text{GBP} = A_v \times f_{3dB}$$

Where:

- Av is the open-loop voltage gain of the amplifier.
- $f3dB$ is the bandwidth of the amplifier, typically defined as the frequency at which the gain drops by 3 dB (half power point)[15].

The GBP represents the frequency at which the amplifier's gain-bandwidth product is constant. Beyond this frequency, the gain starts to decrease due to the internal capacitances and other frequency-dependent effects of the amplifier.

This equation quantifies the trade-off between gain and bandwidth in amplifiers. A higher GBP indicates that the amplifier can maintain its gain over a wider range of frequencies, making it suitable for applications requiring consistent performance across a broad spectrum of signals.

In amplifier design, engineers often optimize the GBP to meet specific performance requirements, such as bandwidth, gain, and stability.

Increasing the GBP typically involves trade-offs between stability, power consumption, and circuit complexity[18].

3.11 Applications

3.11.1 Rectifiers

Rectifiers are electronic circuits used to convert alternating current (AC) into direct current (DC). They play a crucial role in various power supply and signal processing applications, where a steady DC voltage or current is required.

Types of Rectifiers:

1. **Half-Wave Rectifier:**
   - Utilizes a single diode to convert only one half of the AC input waveform into DC.
   - Simplest and most inefficient type of rectifier.

2. **Full-Wave Rectifier:**
   - Utilizes either two or four diodes to convert both halves of the AC input waveform into DC[14,18].
   - More efficient than half-wave rectifiers as it produces a smoother DC output.

3. **Bridge Rectifier:**
   - A type of full-wave rectifier that uses four diodes configured in a bridge arrangement.
   - Provides higher efficiency and better performance compared to other rectifiers.

Working Principle:

In a rectifier circuit, the diode(s) allow current to flow in only one direction, effectively blocking the negative half of the AC waveform. As a result, the output waveform consists of the positive half-cycles of the input

AC voltage, which is then smoothed using a filter capacitor to produce a relatively stable DC voltage.

Mathematical Equations:

The output voltage ($Vout$) of a rectifier circuit can be approximated using the following equations:

1. **Half-Wave Rectifier:** $Vout = Vmax - Vd$ Where:
   - $Vmax$ is the peak value of the input AC voltage.
   - Vd is the forward voltage drop across the diode.

2. **Full-Wave Rectifier:** $Vout = 2 \cdot Vmax - 2 \cdot Vd$ Where:
   - $Vmax$ is the peak value of the input AC voltage.
   - Vd is the forward voltage drop across the diode.

3. **Bridge Rectifier:** $Vout = 2 \cdot Vmax - 2 \cdot Vd$ Where:
   - $Vmax$ is the peak value of the input AC voltage.
   - Vd is the forward voltage drop across the diode.

Applications:
- Rectifiers are widely used in power supplies for electronic devices, such as mobile phones, computers, and televisions.
- They are also used in battery chargers, motor control circuits, and voltage regulators.
- In signal processing applications, rectifiers are used to convert AC signals from sensors and transducers into DC signals for measurement and analysis.

Overall, rectifiers are essential components in electronics and electrical engineering, providing the necessary conversion of AC to DC for a wide range of applications.

3.11.2 Filters

Filters in bipolar junction transistor (BJT) applications are circuits used to selectively pass or attenuate certain frequencies while allowing others to pass through. These filters are commonly employed in audio, communication, and signal processing systems to shape the frequency response of signals. There are various types of filters, including low-pass, high-pass, band-pass, and band-stop filters, each serving different purposes in signal processing[18,19,20].

Types of Filters:

1. **Low-Pass Filter (LPF):**

    - Allows frequencies below a certain cutoff frequency (fc) to pass through while attenuating higher frequencies.

    - Commonly used to remove high-frequency noise from signals or to smooth out abrupt changes in a signal.

2. **High-Pass Filter (HPF):**

    - Allows frequencies above a certain cutoff frequency (fc) to pass through while attenuating lower frequencies.

    - Useful for removing low-frequency noise or DC offset from signals.

3. **Band-Pass Filter (BPF):**

    - Allows frequencies within a specific range (between two cutoff frequencies, $fc1$ and $fc2$) to pass through while attenuating frequencies outside this range.

    - Suitable for extracting specific frequency bands from a signal, such as in radio receivers or audio equalizers.

4. **Band-Stop Filter (BSF) or Notch Filter:**

- Attenuates frequencies within a specific range (between two cutoff frequencies, $fc1$ and $fc2$) while allowing frequencies outside this range to pass through.

- Used to suppress interference or remove unwanted frequency components from signals.

Mathematical Equations:

The transfer function ($H(f)$) of a filter describes its frequency response and can be mathematically represented using various circuit analysis techniques. In general, the transfer function of a filter determines how the output signal ($Vout$) relates to the input signal (Vin) as a function of frequency (f).

1. Low-Pass Filter:

$$H_{LPF}(f) = \frac{1}{1 + \frac{jf}{f_c}}$$

2. High-Pass Filter:

$$H_{HPF}(f) = \frac{jf}{jf + f_c}$$

3. Band-Pass Filter:

$$H_{BPF}(f) = \frac{jf}{(jf + f_{c1})(jf + f_{c2})}$$

4. Band-Stop Filter:

$$H_{BSF}(f) = \frac{(jf + f_{c1})(jf + f_{c2})}{(jf)^2 + (f_{c1} + f_{c2})jf + f_{c1}f_{c2}}$$

3.11.3 Feedback Amplifiers

Feedback amplifiers are circuits where a portion of the output signal is fed back to the input with the purpose of modifying the amplifier's performance. The feedback can be positive or negative, and it affects parameters like gain, bandwidth, input/output impedance, stability, and distortion. Feedback amplifiers are widely used in various applications due to their ability to improve performance and stability.

Types of Feedback:

1. **Negative Feedback (NFB):** In negative feedback, a portion of the output signal is fed back to the input with a phase that opposes the input signal. It reduces gain but improves linearity, bandwidth, and stability.

2. **Positive Feedback (PFB):** In positive feedback, the feedback signal reinforces the input signal, resulting in increased gain but reduced stability. It's less common and mostly used in oscillators and signal generators.

Advantages of Feedback:

1. **Improved Linearity:** Negative feedback reduces distortion and improves linearity.

2. **Increased Bandwidth:** Negative feedback widens the frequency response of the amplifier.

3. **Stability:** Negative feedback increases stability and reduces sensitivity to component variations.

4. **Reduced Distortion:** Negative feedback reduces harmonic distortion.

Feedback Amplifier Equations (Negative Feedback):

For a voltage amplifier with negative feedback, the closed-loop gain (A_{CL}) can be expressed as:

Where:

- A_{OL} is the open-loop gain of the amplifier[1,7,9,14,15].
- β is the feedback factor (the fraction of the output voltage fed back to the input).

The input resistance (Rin) and output resistance ($RoutRout$) of the amplifier with feedback can be approximated as:

$$A_{CL} = \frac{A_{OL}}{1 + \beta A_{OL}}$$

$$R_{in} = \frac{R_{in\_OL}}{1 + \beta A_{OL}}$$

$$R_{out} = \frac{R_{out\_OL}}{1 + \beta A_{OL}}$$

Where:

- Rin_OL and $Rout\_OL$ are the open-loop input and output resistances of the amplifier, respectively.

Feedback Amplifier Equations (Positive Feedback):

Positive feedback amplifiers are less common due to stability issues, but they are characterized by their loop gain (T), which is the gain around the feedback loop. In oscillators, for instance, the condition for sustained oscillation is $|T|=1$.

3.11.4 Oscillators

Oscillators:

Oscillators are electronic circuits that generate periodic waveforms without the need for an external input signal. They are fundamental components in various electronic devices and systems, providing stable and precise timing signals for applications such as clocks, radio frequency (RF) signal generation, and communication systems.

Types of Oscillators[11,14]:

1. **Sinusoidal Oscillators:**

    - Produce sinusoidal waveforms at specific frequencies.

    - Used in audio applications, RF signal generation, and local oscillators in communication systems.

2. **Relaxation Oscillators:**

    - Generate non-sinusoidal waveforms such as square, triangular, or sawtooth waves.

    - Commonly used in pulse-width modulation (PWM) circuits, voltage-controlled oscillators (VCOs), and timing circuits.

3. **Crystal Oscillators:**

    - Utilize the mechanical resonance of a quartz crystal to generate precise and stable oscillations.

    - Widely used in digital clocks, microcontrollers, and frequency synthesizers.

4. **Phase-Locked Loop (PLL) Oscillators:**

    - Consist of a phase detector, voltage-controlled oscillator (VCO), and feedback loop.

    - Used for frequency synthesis, clock recovery, and frequency modulation/demodulation in communication systems[18].

Operating Principles:

Oscillators operate based on the principle of positive feedback, where a portion of the output signal is fed back to the input with the appropriate phase shift to sustain oscillations. The feedback loop continuously reinforces the oscillations, maintaining a stable output waveform[20,21].

Key Components:

1. **Feedback Network:** Determines the frequency and waveform characteristics of the oscillator.

2. **Active Device:** Provides gain and phase shift necessary for sustaining oscillations. Examples include bipolar junction transistors (BJTs), field-effect transistors (FETs), and operational amplifiers (op-amps).

3. **Frequency-Determining Components:** Determine the oscillation frequency and stability. These components can include resistors, capacitors, inductors, and crystals.

Applications:

- Oscillators are essential in electronic devices requiring timing signals, such as clocks, timers, and frequency synthesizers.

- They are used in communication systems for signal generation, modulation, and demodulation.

- Oscillators play a crucial role in test and measurement equipment, audio synthesizers, and signal processing applications[22,23,24].

1. Sinusoidal Oscillator (Wien Bridge Oscillator):

$$f = \frac{1}{2\pi RC}$$

2. Colpitts Oscillator (for LC resonant circuit):

$$f = \frac{1}{2\pi\sqrt{L_1(C_1 C_2)}}$$

3. Crystal Oscillator (for parallel resonant crystal):

$$f = \frac{1}{2\pi\sqrt{LC}}$$

4. Phase-Locked Loop (PLL) Oscillator (VCO frequency):

$$f_{VCO} = f_{ref} \times N$$

3.11.5 Multistage Amplifiers

Multistage amplifiers are circuits composed of two or more amplifier stages cascaded together to achieve higher overall gain, improved bandwidth, and other desired characteristics. These amplifiers are commonly used in various electronic systems, including audio amplifiers, RF amplifiers, and instrumentation amplifiers[15,18,19].

Types of Multistage Amplifiers:

1. **Cascade Amplifiers:** In cascade amplifiers, the output of one amplifier stage is directly connected to the input of the next stage. Each stage provides additional gain and overall amplification.

2. **Cascode Amplifiers:** Cascode amplifiers consist of two or more amplifier stages connected in series, with the output of one stage connected to the input of the next stage through a common emitter or common source configuration.

3. **Darlington Pair:** The Darlington pair is a special configuration where two bipolar junction transistors (BJTs) are connected in a cascaded configuration to achieve high current gain.

4. **Operational Amplifier (Op-Amp) Circuits:** Op-amps can be configured in multistage amplifier circuits to achieve desired gain, impedance matching, and other characteristics.

Operating Principles:

Multistage amplifiers operate based on the principle of signal amplification through successive stages. Each amplifier stage provides gain and helps in shaping the overall frequency response of the amplifier. The input signal is sequentially amplified as it passes through each stage, resulting in higher overall gain[22].

Key Components:

1. **Amplification Stages:** Each stage typically consists of an active device such as a transistor or operational amplifier, along with associated passive components like resistors, capacitors, and inductors.

2. **Coupling Networks:** Coupling capacitors or transformers are used to couple the output of one stage to the input of the next stage while blocking DC bias voltages.

3. **Biasing Circuits:** Biasing circuits are employed to set the operating point of the active devices within each amplifier stage for optimal performance.

Equations for Multistage Amplifiers:

1. **Overall Voltage Gain (*Av*):** The overall voltage gain of a multistage amplifier is the product of the voltage gains of each individual stage. $Av = Av1 \times Av2 \times ... \times Avn$

2. **Overall Current Gain (*Ai*):** Similarly, the overall current gain of a multistage amplifier is the product of the current gains of each individual stage.

3. **Overall Power Gain (*Ap*):** The overall power gain of a multistage amplifier is the product of the power gains of each individual stage[17].

Applications:

Multistage amplifiers find applications in various fields such as audio amplification, RF communication systems, instrumentation, and control systems where high gain, bandwidth, and performance are required.

Overall Voltage Gain (*Av*): $A_v = A_{v1} \times A_{v2} \times ... \times A_{vn}$

Overall Current Gain (*Ai*): $A_i = A_{i1} \times A_{i2} \times ... \times A_{in}$

Overall Power Gain (*Ap*): $A_p = A_{p1} \times A_{p2} \times ... \times A_{pn}$

Voltage Gain of a Single Amplifier Stage (*Avi*): $A_{vi} = \frac{v_{out}}{v_{in}}$

Current Gain of a Single Amplifier Stage (*Aii*): $A_{ii} = \frac{i_{out}}{i_{in}}$

Power Gain of a Single Amplifier Stage (*Api*): $A_{pi} = \frac{P_{out}}{P_{in}}$

Input Impedance of a Single Amplifier Stage (*Zin*): $Z_{in} = \frac{v_{in}}{i_{in}}$

Output Impedance of a Single Amplifier Stage (*Zout*): $Z_{out} = \frac{v_{out}}{i_{out}}$

Chapter-4

Field Effect Transistors/ FETs

4.1 Introduction to MOS structure as two terminal capacitors

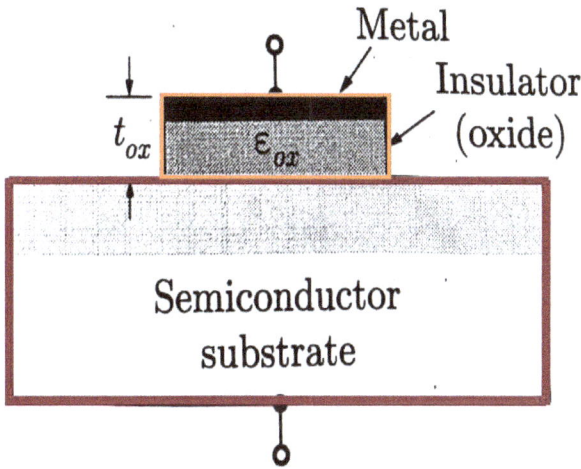

Figure 15 Basic MOS Capacitor structure.

1. **Structure**: The MOS capacitor consists of a metal gate electrode separated from the semiconductor substrate by a thin insulating oxide layer. When a voltage is applied to the metal gate electrode with respect to the semiconductor substrate, an electric field is created across the oxide layer. This field induces a depletion region in the semiconductor substrate beneath the oxide layer, resulting in the formation of a capacitor-like structure.

2. **Operation**: The MOS capacitor operates based on the principle of a parallel plate capacitor. When a voltage V_{gs} Vgs (gate-to-source voltage) is applied to the gate electrode, it creates an electric field across the insulating oxide layer. This field induces a charge in the

semiconductor substrate, causing the formation of depletion regions near the oxide-semiconductor interface. The capacitance of the MOS structure depends on the thickness of the oxide layer and the area of the capacitor plates (gate and substrate).

3. **Examples**:

    - MOSFET (Metal-Oxide-Semiconductor Field-Effect Transistor): MOSFETs are one of the most widely used semiconductor devices and are based on the MOS structure. They consist of a MOS capacitor (gate oxide, metal gate electrode, and semiconductor substrate) with an additional source and drain terminals. The application of a gate-to-source voltage modulates the conductivity of the semiconductor channel between the source and drain terminals, allowing for amplification and switching functions.

    - MOS Capacitor[12,18,22]: MOS capacitors are used in various analog and digital integrated circuits for applications such as decoupling, filtering, and signal processing. They provide capacitance with precise control over parameters such as capacitance value and voltage dependency, making them essential components in semiconductor device fabrication.

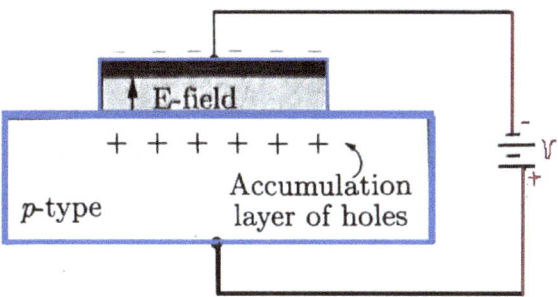

Figure 16 Consideration of MOS as cap with E.

Capacitance of MOS Capacitor: $C_{MOS} = \frac{\varepsilon_{ox} \cdot A}{t_{ox}}$

Gate Capacitance: $C_{gate} = \frac{\varepsilon_{ox} \cdot A_{gate}}{t_{ox}}$

where:

- ε_{ox} is the permittivity of the oxide layer,
- A is the area of the capacitor plates (gate and substrate), and
- t_{ox} is the thickness of the oxide layer.
- A_{gate} is the area of the gate electrode.

4.2.1 Energy Band for MOS Capacitors with p-type substrate

In MOS capacitors with a p-type substrate, the energy band diagram depicts the distribution of energy levels within the semiconductor material[15,18,22].

The energy band diagram shows the valence band (VB), conduction band (CB), and Fermi level (EF) relative to the semiconductor material's energy levels.

P-Type Substrate: In a p-type substrate, majority carriers are holes (positive charge carriers), while minority carriers are electrons.

The Fermi level (EF) lies closer to the valence band (VB) due to the predominance of holes.

Depletion Region: When a negative voltage is applied to the gate electrode with respect to the p-type substrate, it induces an electric field across the oxide layer.

This field causes a depletion region to form in the semiconductor substrate beneath the oxide layer.

In the energy band diagram, the depletion region is depicted as a region where the energy levels bend downwards due to the presence of the electric field[12,24,25,26].

Threshold Voltage: The threshold voltage (Vth) is the voltage required to create a strong inversion layer in the semiconductor substrate, allowing for the formation of a conductive channel between the source and drain terminals[11,13,17].

In the energy band diagram, the threshold voltage corresponds to the point where the energy levels at the semiconductor surface align with the bottom of the conduction band.

Capacitance Behavior: The capacitance of the MOS capacitor depends on the density of states in the semiconductor substrate and the voltage applied to the gate electrode.

At low voltages (below threshold), the capacitance is primarily determined by the depletion region's width and the doping concentration of the substrate.

At high voltages (above threshold), the capacitance is influenced by the formation of an inversion layer and the increased density of charge carriers near the semiconductor surface.

Applications: MOS capacitors with p-type substrates are widely used in integrated circuits for various applications, including memory devices, analog circuits, and digital circuits.

Understanding the energy band diagram and capacitance behavior is essential for designing and optimizing MOS capacitors for specific applications[18,19,20,25,26].

Energy Band Diagram: The energy band diagram ($E(x)$) for a MOS capacitor with a p-type substrate can be represented as:

$$E(x) = E_{CB}(x) - E_{VB}(x)$$

where $E_{CB}(x)$ is the energy level of the conduction band and $E_{VB}(x)$ is the energy level of the valence band as a function of position xx within the semiconductor material.

Fermi-Dirac Distribution: The Fermi-Dirac distribution function ($f(E)$) describes the probability of occupying an energy level EE in a semiconductor material at thermal equilibrium and can be expressed as:

$$f(E) = \frac{1}{1 + e^{\frac{E-E_F}{kT}}}$$

where E_F is the Fermi level, k is Boltzmann's constant, and T is the temperature in Kelvin.

Depletion Region Width: The width (W_d) of the depletion region in the semiconductor substrate can be calculated using the Poisson equation:

$$\frac{d^2\phi}{dx^2} = -\frac{\rho}{\varepsilon_s}$$

where ϕ is the electric potential, ρ is the charge density, and ε_s is the permittivity of the semiconductor.

Threshold Voltage: The threshold voltage (Vth) of the MOS capacitor can be determined by equating the charge in the depletion region to the charge induced by the gate voltage, given by:

$$Q_{dep} = -Q_{inv}$$

where Qdep is the depletion region charge and Qinv is the inversion layer charge.

Capacitance Behavior: The capacitance (C) of the MOS capacitor is related to the charge density (ρ) in the depletion region and the permittivity (εox) of the oxide layer by the equation[11,14]:

$$C = \frac{\varepsilon_{ox}}{W_d}$$

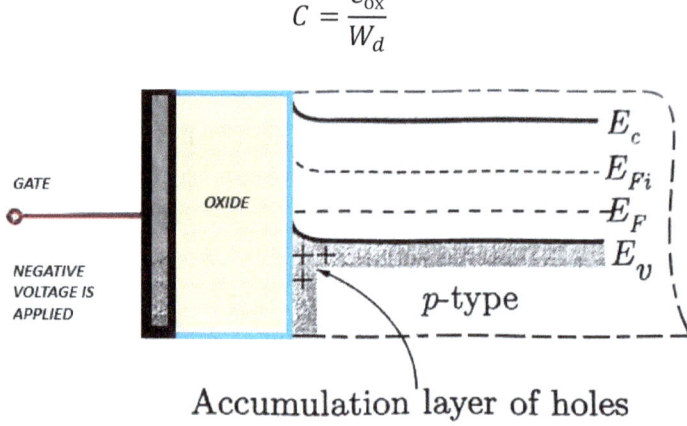

Figure 17 Energy band diagram for MOS CAPACITOR for P- substrate for negative bias.

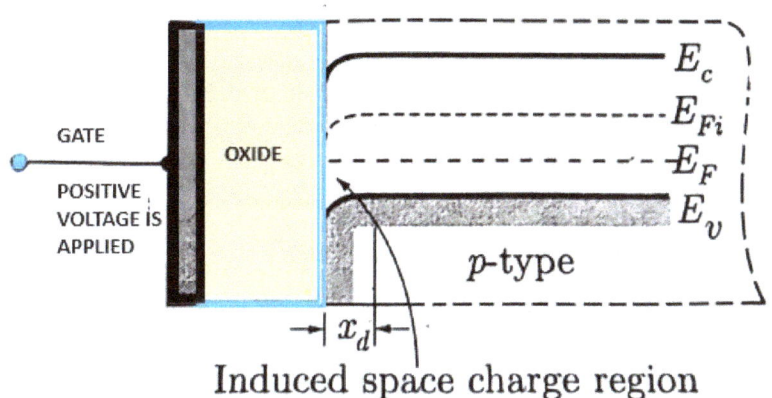

Figure 18 Energy band diagram for MOS CAPACITOR for moderate positive bias.

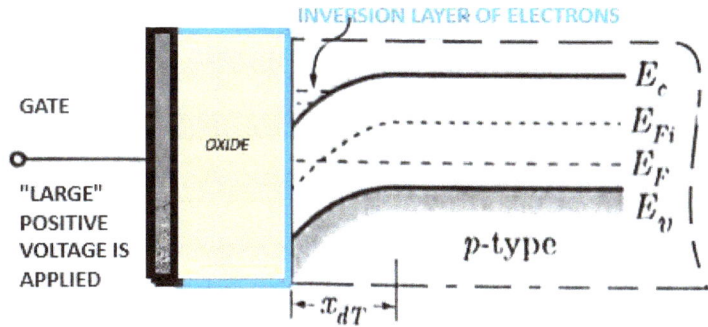

Figure 19 Energy band diagram for MOS CAPACITOR for large positive bias.

4.2.2 Energy Band for MOS Capacitors with n-type substrate

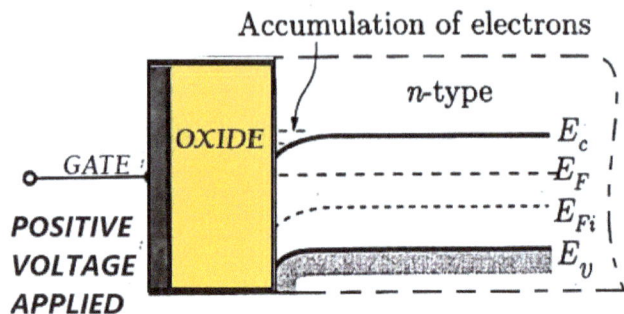

Figure 20 Energy band diagram for (N TYPE SUBSTRATE) MOS CAPACITOR for positive GATE bias.

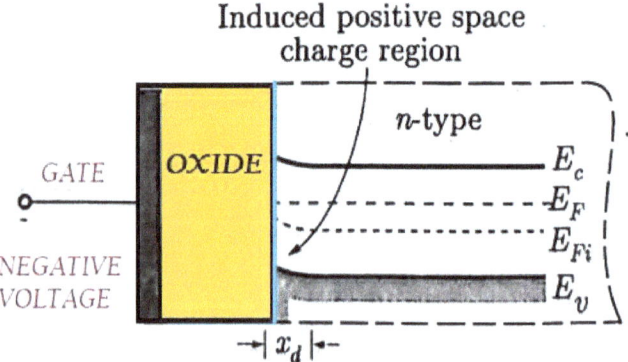

Figure 21 Energy band diagram for (N TYPE SUBSTRATE) MOS CAPACITOR for moderate negative GATE bias.

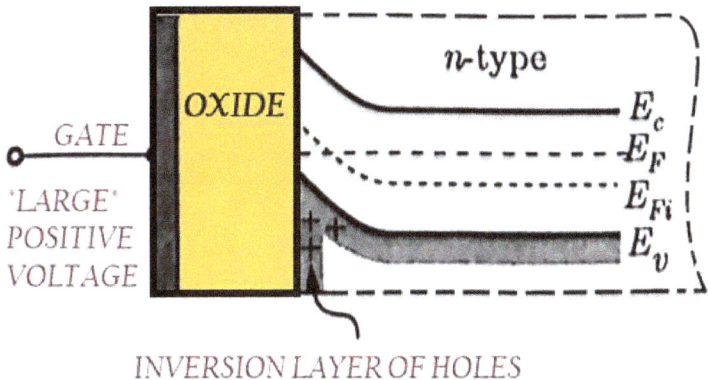

Figure 22 Energy band diagram for (N TYPE SUBSTRATE) MOS CAPACITOR for LARGE negative GATE bias.

4.3 Types of FET and symbology.

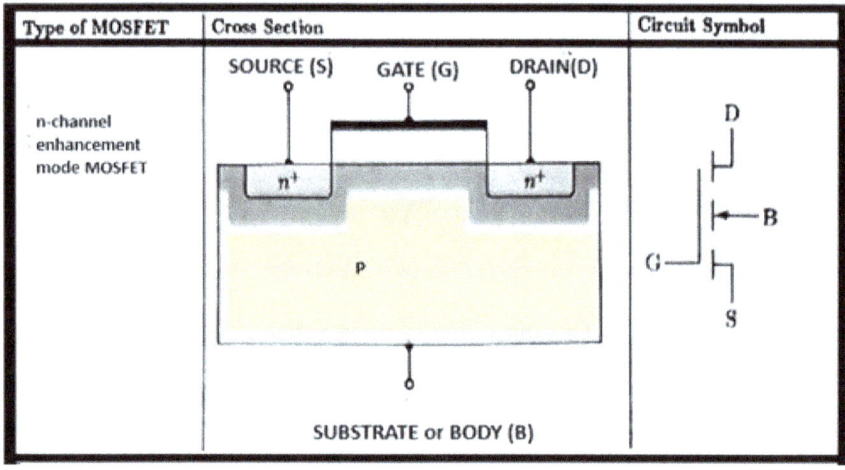

(A)

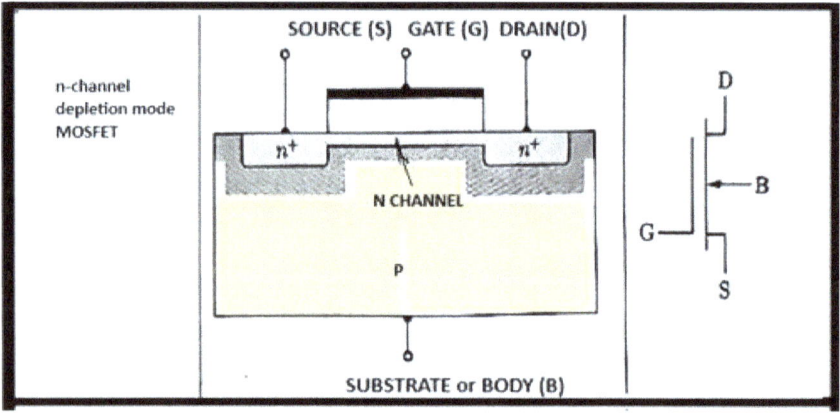

(B)

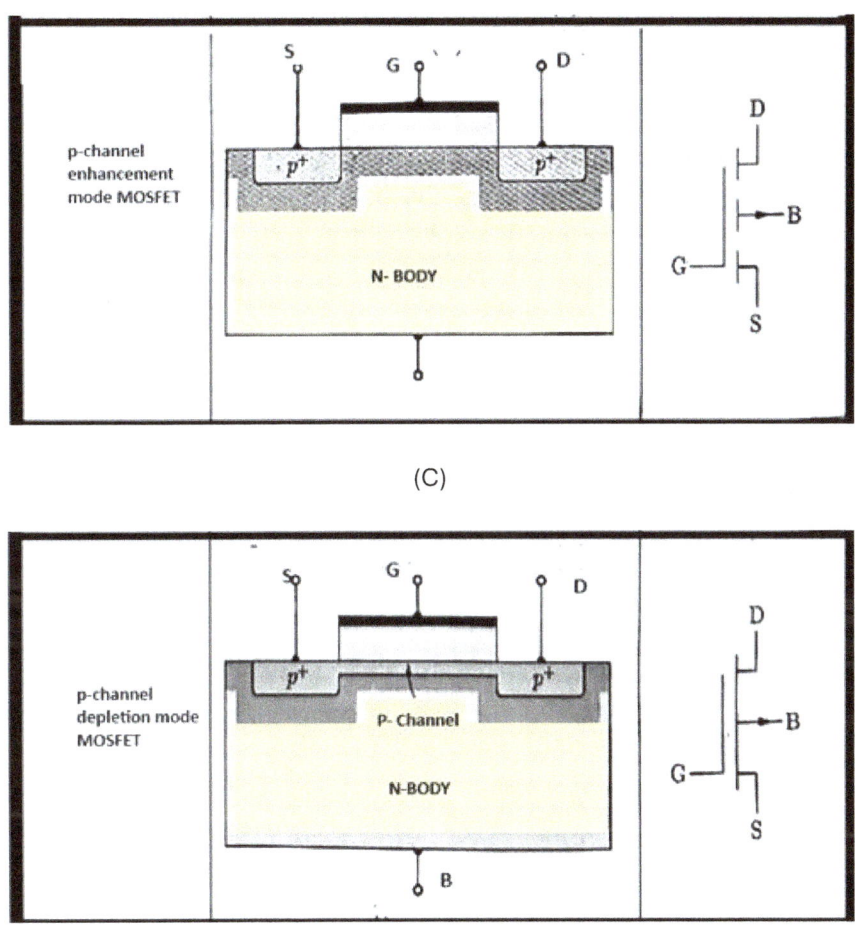

(C)

(D)

Figure 23 Types of FET

4.4 MOSFET Biasing and Configuration

Biasing is the process of setting the DC[11,18] operating point, or quiescent point (Q-point), of a transistor circuit. In MOSFET circuits, biasing ensures that the transistor operates in the desired region and exhibits the desired characteristics. The biasing conditions determine the operating point of the MOSFET and affect its behavior in different operating regions[24].

1. Operating Regions of MOSFET:

    - Cut-off Region: In this region, both the gate-to-source voltage (VGS) and the drain-to-source voltage (VDS) are below the threshold voltage (Vth). As a result, the MOSFET is turned off, and no current flows between the drain and source ($ID=0$).

    - Triode (or Linear) Region: In this region, the gate-to-source voltage (VGS) is greater than the threshold voltage (Vth), but the drain-to-source voltage (VDS) is less than the gate-to-source voltage minus the threshold voltage ($VGS-Vth$). The MOSFET operates as a voltage-controlled resistor, and the drain current (ID) is proportional to the gate-to-source voltage (VGS).

    - Saturation Region: In this region, both the gate-to-source voltage (VGS) and the drain-to-source voltage (VDS) are sufficient to keep the MOSFET fully turned on. The MOSFET operates as a current source, and the drain current (ID) remains relatively constant with variations in the drain-to-source voltage (VDS).

2. Biasing Techniques:

    - Fixed Bias: In this technique, a DC voltage source is directly connected to the MOSFET gate terminal to establish the desired biasing conditions.

    - Self-Bias (or Source Bias): In this technique, a resistor is connected between the gate terminal and the source terminal of the MOSFET. The voltage drop across the resistor biases the MOSFET and stabilizes the Q-point.

- Voltage Divider Bias: In this technique, a voltage divider network consisting of resistors is used to bias the MOSFET. The midpoint of the voltage divider provides the bias voltage for the gate terminal.

- Cascode Bias: In this technique, two MOSFETs are connected in series, with the gate of the lower MOSFET connected to the drain of the upper MOSFET. This configuration enhances stability and reduces sensitivity to variations in the supply voltage.

3. Bias Stability:

- Bias stability refers to the ability of the biasing circuit to maintain the Q-point of the MOSFET relatively constant despite variations in temperature, transistor parameters, and power supply voltage.

- Proper biasing ensures that the MOSFET operates in the desired region under all operating conditions, thereby optimizing performance and minimizing distortion.

4.4.1 Models and Equations with Characterization

MOSFET Operating Regions:

- Cut-off Region: VGS<Vth and $VDS<VGS-Vth$

- Triode (Linear) Region: VGS>Vth and $0<VDS<VGS-Vth$

- Saturation Region: $VGS>Vth$ and $VDS>VGS-Vth$

Drain Current Equations: Triode Region:

$$I_D = \frac{1}{2}\mu_n C_{ox} \frac{W}{L}(V_{GS} - V_{th})^2$$

Saturation Region:

$$I_D = \frac{1}{2}\mu_n C_{ox} \frac{W}{L}(V_{GS} - V_{th})^2(1 + \lambda V_{DS})$$

Biasing Equations:

$$Fixed\ Bias: V_{GS} = V_{GG}$$

$$Self\text{-}Bias: V_{GS} = V_{GG} - I_D R_S$$

$$Voltage\ Divider\ Bias: V_{GS} = \frac{R_2}{R_1 + R_2} V_{DD}$$

$$Cascode\ Bias: V_{GS2} = V_{DD} - I_D R_{D1} - I_D R_{S2}$$

Small Signal Parameters:

$$Transconductance\ g_m = \frac{\partial I_D}{\partial V_{GS}}\bigg|_{Q-point}$$

$$Output\ Conductance: g_d = \frac{\partial I_D}{\partial V_{DS}}\bigg|_{Q-point}$$

$$C_{gd} = \frac{\partial Q_{gd}}{\partial V_{gd}},\quad C_{gs} = \frac{\partial Q_{gs}}{\partial V_{gs}},\quad C_{ds} = \frac{\partial Q_{ds}}{\partial V_{ds}}$$

4.5 Comparison between FINFET, MOSFET, Replacement Gate, Floating Gate

Floating Gate Transistor (FGT) and FinFET (Fin Field-Effect Transistor) are both advanced transistor technologies, but they serve different purposes and have different structures:

Floating Gate Transistor (FGT):

- **Purpose:** FGT is commonly used in non-volatile memory devices, particularly in Flash memory[25].

- **Structure:** In FGT, a conductive floating gate is electrically isolated by a thin insulating layer. Charge can be injected onto or removed from the floating gate using quantum tunneling mechanisms, allowing the device to store binary data.

- **Applications:** Flash memory, EEPROM (Electrically Erasable Programmable Read-Only Memory).

FinFET:

- **Purpose:** FinFET is a type of MOSFET designed to overcome limitations of traditional planar MOSFETs, particularly in terms of power consumption and leakage current[1,2,5,8,9].

- **Structure:** In FinFET, the channel region is formed on a thin silicon fin protruding from the substrate, with the gate wrapped around three sides of the fin. This three-dimensional structure helps to improve electrostatic control and reduces leakage current.

- **Applications:** Digital logic circuits, microprocessors, system-on-chip (SoC) designs, and other high-performance integrated circuits.

On the other hand, MOSFET (Metal-Oxide-Semiconductor Field-Effect Transistor) and Replacement Gate Transistor are different types of transistors with distinct characteristics:

MOSFET:

- **Purpose:** MOSFETs are fundamental building blocks of modern integrated circuits, used in a wide range of applications including digital and analog circuits.

- **Structure:** In a MOSFET, a metal gate electrode is separated from the semiconductor channel by a thin insulating layer (oxide), typically made of silicon dioxide (SiO_2). The gate voltage controls the flow of current through the channel.

- **Variants:** Various types of MOSFETs exist, including planar MOSFETs, FinFETs, and others, each with different structures and characteristics.

Replacement Gate Transistor:

- **Purpose:** Replacement Gate Transistor is a type of transistor used in advanced semiconductor manufacturing processes, particularly in the fabrication of FinFETs.

- **Structure:** In Replacement Gate Technology, the gate material is deposited after the formation of the fin structure, allowing for more precise control over gate dimensions and reducing variability in gate lengths.

- **Advantages:** Replacement Gate Technology enables tighter control over transistor dimensions, improving transistor performance, power efficiency, and yield in advanced semiconductor processes[11,14,18].

4.5.1 Mathematical equations

Threshold Voltage Equation: The threshold voltage (Vth) of a FinFET can be calculated using the following equation:

$$V_{th} = V_{FB} + \gamma(\Phi_s + 2\phi_F - 2qN_a\epsilon_{si} - \Phi_s + 2\phi_F)$$

Transconductance Equation: The transconductance ($gmgm$) of a Replacement Gate Transistor can be expressed as:

$$g_m = \mu_n C_{ox} \left(\frac{W}{L}\right) V_{GS}$$

Charge Injection Equation: The charge (Qf) stored in the floating gate of a Floating Gate Transistor can be calculated using the charge injection equation[22,24,25,26]:

$$Q_f = Q_c + Q_t$$

where:

- V_{FB} is the flat-band voltage,
- γ is the body factor,
- Φ_s is the surface potential,
- ϕ_F is the Fermi potential,
- Na is the acceptor doping concentration,
- ε_{si} is the permittivity of silicon.
- μ_n is the electron mobility,
- Cox is the gate oxide capacitance per unit area,
- W_L is the aspect ratio of the transistor,
- V_{GS} is the gate-source voltage.

Charge Retention Equation:

$$\frac{dQ_f}{dt} = -I_{leakage}$$

Program/Erase Voltage Equation:

$$V_{PE} = f(programming/erasing\ mechanism)$$

Replacement Gate Transistor:

$$I_{DS} = f(V_{GS}, V_{DS}, W/L, \mu_n, C_{ox})$$

Drain Current Equation:

$$I_{DS} = f(V_{GS}, V_{DS}, W, L, V_{th}, process\ variations)$$

Subthreshold Slope Equation: $\frac{dV_{GS}}{d(log(I_{DS}))}$

FinFET: Drain-Source Current Equation:

$$I_{DS} = f(V_{GS}, V_{DS}, W, L, V_{th}, process\ variations)$$

Channel Length Modulation Equation: $\lambda = \frac{\partial I_{DS}}{\partial V_{DS}}$

4.5.2 Capacitance-Voltage characteristics of a MOS Cap

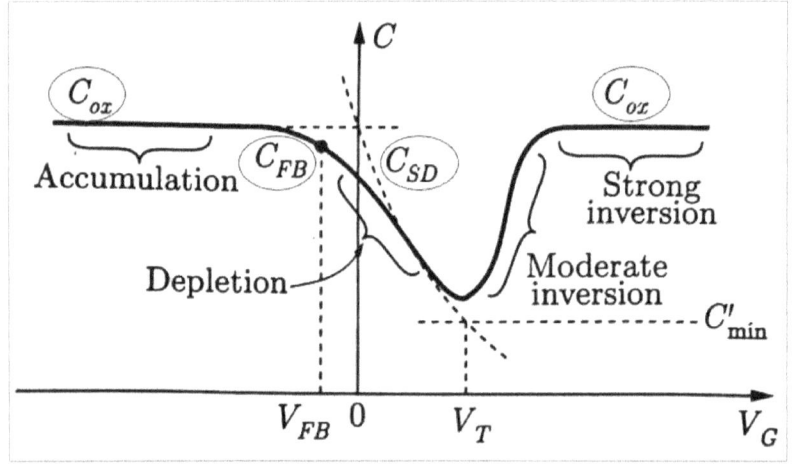

Figure 24 CAPACITANCE-VOLTAGE characteristics of a MOS

4.6 Second order Effects in FETs

Second channel effects, also known as short-channel[11,14,18,19,20,25,26] effects, refer to the phenomenon that occurs in short-channel MOSFETs, where the channel length is comparable to or smaller than the depletion layer width. These effects become significant as semiconductor device dimensions shrink, leading to challenges in maintaining device performance and reliability. Some common second channel effects include:

1. **Threshold Voltage Roll-Off:** Short-channel MOSFETs exhibit a decrease in threshold voltage (V_th) as the channel length decreases. This roll-off occurs due to the increased influence of the source and drain regions on the channel, reducing the effective gate control.

2. **Drain-Induced Barrier Lowering (DIBL):** DIBL refers to the reduction in the barrier height between the source and drain regions as the drain voltage increases. In short-channel devices, the electric field from the drain extends deeper into the channel, causing a decrease in the effective barrier height and hence a decrease in V_th.

3. **Subthreshold Slope Degradation:** The subthreshold slope (SS), which measures the sensitivity of drain current (I_D) to changes in gate voltage (V_G) in the subthreshold region, worsens in short-channel devices. This degradation is due to increased drain-induced barrier lowering and increased leakage currents.

4. **Hot Carrier Injection:** Short-channel MOSFETs are more susceptible to hot carrier injection, where carriers gain enough energy to become trapped in the gate oxide. This phenomenon can degrade device reliability and increase leakage currents.

5. **Channel Length Modulation (Early Effect):** In short-channel devices, the drain-induced electric field extends further into the

channel, causing carriers to gain velocity before reaching the drain. This effect leads to a decrease in the effective channel length and an increase in drain current with drain voltage[15,18,19].

Equations:

4.6.1 Threshold Voltage Roll-Off

The threshold voltage roll-off (ΔV_th) in short-channel MOSFETs can be approximated using the following equation

$$\Delta V_{th} = \frac{\sqrt{2q\epsilon_s N_a}}{C_{ox}} \cdot \left(\frac{1}{\sqrt{\phi_F + V_{SB}}} - \frac{1}{\sqrt{\phi_F}} \right)$$

4.6.2 Drain-Induced Barrier Lowering (DIBL)

DIBL can be quantified by the change in threshold voltage ($\Delta V th(DIBL)$) due to drain voltage (VDD) using an empirical equation such as[1,2,5]:

$\Delta V_{th(DIBL)} = k \cdot V_{DD}$ where k is a process-dependent constant.

4.6.3 Hot Carrier Injection

Hot carrier injection can be characterized by the change in threshold voltage ($\Delta V th(HCI)$) due to gate voltage (VGS) using an empirical equation such as:

$\Delta V_{th(HCI)} = \alpha \cdot V_{GS}$ where α is a process-dependent constant.

4.6.4 Subthreshold Slope Degradation

The subthreshold slope (SS) can be calculated using the following equation, which accounts for DIBL effects:

$$SS = \frac{dV_G}{d(\ln(I_D))} = \frac{1}{C_{ox}}\left(\frac{2kT}{q} \cdot \frac{1}{1+\frac{C_{dep}}{C_{ox}}} + \frac{V_{DS}+V_{DD}}{V_{T0}}\right)$$

- q is the elementary charge,
- ϵ_s is the permittivity of the semiconductor,
- Na is the acceptor doping concentration,
- C_{ox} is the gate oxide capacitance per unit area,
- ϕF is the Fermi potential,
- V_{SB} is the source-to-bulk voltage.
- k is Boltzmann's constant,
- T is the absolute temperature,
- C_{dep} is the depletion layer capacitance,
- V_{DS} is the drain-to-source voltage,
- V_{DD} is the drain voltage, and
- V_{T0} is the threshold voltage.

Index

a

a reverse bias voltage, 30, 34, 42
admittance, 5, 39, 40
advanced, 2, 55, 83, 84
alpha, 54
ambipolar transport, 4, 27, 28
amplification, 25, 49, 51, 57, 67, 74, 75, 76

b

band-pass, 70, 71
band-stop, 70, 71
base transit, 65
beta, 54
bias stability, 82
biasing, 6, 7, 51, 52, 53, 57, 74, 82, 83
boltzmann's constant, 31, 55, 78, 88

c

capacitance, 5, 32, 35, 36, 38, 39, 45, 46, 60, 61, 64, 66, 67, 76, 78, 79, 85, 87, 88
capacitance, 4, 5, 7, 32, 35, 36, 38, 45, 46, 77, 78, 79, 86
cascade, 74
cascode, 74, 82, 83
channel length modulation, 86
characteristic, 8, 43, 44, 46
charge, 8, 9, 10, 11, 12, 14, 16, 17, 18, 19, 20, 21, 22, 25, 28, 29, 30, 31, 32, 33, 35, 36, 37, 38, 46, 48, 49, 50, 55, 76, 77, 78, 79, 85, 87
charge injection equation, 85
colpitts oscillator, 73
common base, 51
common collector, 52
common emitter, 51
compensated semiconductor, 11
compensation, 62
conductivity, 9, 10, 11, 22, 26, 34, 45, 48, 67, 76
configuration, 51, 52, 58, 59, 68, 74, 82
configuration, 6, 7, 51, 52, 58, 59, 82
crystal, 10, 11, 26, 73
crystal oscillator, 73

d

darlington, 74
defect, 3, 26, 37
depletion region width, 78
design and simulation, 6, 59
diffusion length, 23
diode resistance, 4, 32
dopant, 11
drift and diffusion, 53
drift current density, 19

e

early effect, 87
early voltage, 55, 61
ebers-moll, 6, 55, 56
electric field, 4, 8, 16, 17, 18
electrical characteristics, 11, 50
electronics, 2, 9, 48, 70
electrons, 26, 36, 49
emitter current, 54, 55
energy band, 9, 10, 42, 77, 78
energy band, 4, 7, 8, 10, 12, 13, 77, 78, 80
equation, 9, 10, 17, 23, 24, 29, 32, 33, 34, 35, 37, 41, 54, 55, 66, 67, 68, 78, 84, 87
equation, 4, 14, 16, 20, 21, 23, 24, 29, 32, 33, 53, 54, 55, 84, 85, 86
equilibrium, 4, 19, 20, 29, 37
equivalent circuit, 8, 45, 58, 59
extrinsic, 9

f

feedback, 6, 60, 71, 72, 73
fermi level, 4, 12, 13, 15, 42
fermi-dirac, 14, 15, 16, 78
field-effect, 6, 48, 49, 64, 76, 83, 84
filters, 6, 46, 70
floating gate, 7, 83, 85
force, 16, 20, 28, 29
forward bias, 5, 8, 36, 38, 41, 42

frequency, 6, 17, 22, 39, 44, 46, 47, 52, 56, 57, 59, 60, 61, 62, 63, 64, 65, 66, 67, 68, 70, 71, 73, 74

g

gain-bandwidth product, 68
gummel-poon, 6, 56, 60, 61

h

half-wave rectifier, 69
hall voltage, 29
heating, 43, 61
heavy doping, 11
high-speed switches, 44
hole, 9, 10, 11, 12, 14, 19, 21, 22, 23, 27, 28, 31, 37, 43

i

impedance, 49, 57, 75
intrinsic, 9, 22, 63, 91
iv characteristics, 8, 40, 41

j

jfet, 48
junction, 5, 6, 35, 36, 42, 45, 48, 49, 50, 53

l

lattice, 10, 11, 17, 22, 26, 27, 43
light doping, 11

m

mean free path, 16, 19
mean free path, 4, 16
metals, 13, 15, 22
mobility, 4, 9, 10, 11, 12, 16, 17, 19, 21, 22, 24, 26, 28, 85
model, 4, 6, 9, 19, 55, 56, 58, 59, 61, 63, 64
models, 39, 56, 57, 58, 59, 61, 64, 67
mos, 6, 7, 8, 76, 77, 78, 79, 80, 86
mosfet, 7, 48, 63, 64, 76, 82, 83, 84
multistage, 6, 74, 75

n

n-type, 4, 7, 9, 10, 11, 13, 16, 19, 22, 23, 25, 35, 42, 43, 45, 48, 49, 50, 80

o

operation, 6, 46, 48, 52, 53, 76
operational, 74
oscillators, 6, 44, 46, 72, 73

p

parameter, 6, 56, 65
permittivity, 33
phase-locked loop, 73, 74
phosphorus, 10
pn junction diode, 30, 32, 36
potential, 10, 15, 25, 29, 30, 31, 33, 34, 35, 37, 61, 62, 78, 85, 87
power gain, 75
power gain, 75
probability, 14, 15, 92
program/erase, 85
p-type, 4, 7, 9, 10, 11, 13, 16, 19, 22, 23, 25, 35, 42, 43, 45, 48, 49, 50, 77, 78

r

radio frequency, 45, 46, 72
real-world applications, 6, 61
rectifiers, 6, 45, 68, 69
resistivity, 4, 21, 22
roll-off, 86

s

saturation, 17, 52, 53, 55, 82, 83
saturation velocity, 17
scattering, 4, 22, 26
schottky diode, 5, 46
schrödinger, 10
second channel effects, 86
shockley diode, 4, 32, 55
signal, 36, 39, 40, 46, 48, 51, 52, 56, 57, 58, 59, 63, 64, 65, 67, 68, 69, 70, 71, 72, 73, 74, 76
sinusoidal oscillator, 73

[127]

small signal parameters, 83
small-signal, 57, 58, 59, 63, 64, 65
spontaneous emission, 27
structure, 6, 30, 49, 64, 76, 83, 84
subthreshold slope equation, 85
susceptance, 39
switching times, 45

transistors, 23, 24, 28, 42, 48, 51, 57, 58, 62, 64, 73, 74, 84
transistors, 5, 6, 25, 28, 48, 57, 62, 76
transit time effects, 62
transition capacitance, 36
triode, 82, 83
troubleshooting, 6, 61
tunnel diode, 8, 43, 44

t

temperature, 4, 13, 14, 15, 16, 17, 19, 20, 22, 23, 24, 44, 47, 55, 65, 66
threshold, 88
transistor, 5, 6, 47, 48, 49, 50, 51, 52, 53, 54, 64, 76, 83, 84, 85

v

validity, 6, 65

References

1. Naveen Kumar Macha, Bhavana Tejaswini Repalle, Md Arif Iqbal, Mostafizur Rahman, "Crosstalk-Computing-Based Gate-Level Reconfigurable Circuits", *IEEE Transactions on Very Large Scale Integration (VLSI) Systems*, vol.30, no.8, pp.1073-1083, 2022.
2. Waynant, Ronald W., and John K. Lowell. *Electronic and Photonic Circuits and Devices*. 1999.
3. Patrick, Dale R., Stephen W. Fardo, Ray E. Richardson, and Vigyan (Vigs) Chandra. *Electronic Devices and Circuit Fundamentals*. 2022. River Publishers.
4. Zhang, Bo, and Dongyuan Qiu. *Sneak Circuits of Power Electronic Converters*. 2014. Wiley-IEEE Press.
5. J. Langeheine, K. Meier, J. Schemmel, "Intrinsic evolution of quasi DC solutions for transistor level analog electronic circuits using a CMOS FPTA chip", *Proceedings 2002 NASA/DoD Conference on Evolvable Hardware*, pp.75-84, 2002.
6. Higuchi T. et al. Evolvable Hardware with Genetic Learning: A first step towards building a Darwin machine. In J-A. Meyer, H. L. Roitblat and S. W. Wilson (eds.), Proceedings of the 2 International Conference on the Simulation of Adaptive Behavior, pp 417-424, 1992. MIT Press: Cambridge
7. Ning, Z-Q, Mouthaan, T. and Wallinga, H. SEAS: A Simulated Evolution Approach for Analog Circuit Synthesis. In Proceedings of the IEEE 1991 Custom Integrated Circuits Conference, pp 5.2.1-5.2.4. May 12-15, 1991, San Diego, USA. IEEE Press: Piscataway, NJ.
8. Gontrand, Christian. *Analog Devices and Circuits*. 2023. Wiley Telecom.
9. Ibe, Eishi H. *Terrestrial Radiation Effects in ULSI Devices and Electronic Systems*. 2015. Wiley-IEEE Press.
10. Tripathi, Suman Lata, and Smrity Dwivedi, eds. *Electronic Devices and Circuit Design: Challenges and Applications in the Internet of Things*. 1st ed., February 2022.
11. Basic Engineering Circuit Analysis, 6/e, J. David Irwin and Chwan-Hwa Wu, John Wiley & Sons, New York, 1999, ISBN 0-471-36574-2
12. Basic Engineering Circuit Analysis, 7/e, J. David Irwin, John Wiley & Sons, New York, 2002, ISBN 0-471-40740-2

13. Boylestad, Robert L., and Louis Nashelsky. Electronic Devices and Circuit Theory. 10th ed. Paperback, January 1, 2009.
14. Paynter, Robert T. *Introductory Electronic Devices And Circuits: Conventional Flow Version*. 7th ed. January 1, 2005.
15. Gupta, J.B. *Electronic Devices and Circuits*. Paperback, January 1, 2012.
16. Patil, Mahesh B. *Basic Electronic Devices and Circuits*. Kindle Edition, July 18, 2013.
17. Raju, G S N. Electronic Devices and Circuits. Paperback, September 19, 2006.
18. Sze, S. M., Yiming Li, and Kwok K. Ng. *Physics of Semiconductor Devices*. 4th ed., March 3, 2021.
19. Casey, H. Craig. *Devices for Integrated Circuits: Silicon and III-V Compound Semiconductors*. 1st ed., December 14, 1998.
20. Whitaker, Jerry C. *Semiconductor Devices And Circuits*. 7th ed. Electronics Handbook Series. Hardcover, Import, December 29, 1999.
21. Roberts, Jonathan. *Using Imperfect Semiconductor Systems for Unique Identification*. 1st ed., 2017 Edition. Springer Theses. Kindle Edition. Springer, 2017.
22. Neamen, Donald A. *Semiconductor Physics And Devices: Basic Principles*. 4th ed., January 2011.
23. Sedra, Adel S., Kenneth C. (KC) Smith, Tony Chan Carusone, and Vincent Gaudet. *Microelectronic Circuits*. 8th ed. The Oxford Series in Electrical and Computer Engineering. November 15, 2019.
24. Gubner, John A. *Probability and Stochastic Processes: A Friendly Introduction for Electrical and Computer Engineers*. 3rd ed. January 28, 2014.
25. Spahic, Benjamin. *Electrical Engineering Without Prior Knowledge: Understand the Basics Within 7 Days (Become an Engineer Without Prior Knowledge)*. October 23, 2020.
26. Nenni, Daniel, and Paul McLellan. *Fabless: The Transformation of the Semiconductor Industry*. April 1, 2014.